The Fascinating World of Radio Communications

Edited By Wayne Green

FIRST EDITION

FIRST PRINTING—OCTOBER 1971

Printed in the United States
of America

International Standard Book No. 0-8306-1586-5

Library of Congress Card Number: 79-178693

Preface

Radio Communications, either one way or two way, can be a whole world in itself. Millions of people around the world enjoy listening to short wave broadcasting stations, about 400,000 or so have amateur radio stations so they can talk with each other, another half million or so are active on the American citizens band with two-way radios, and untold thousands are having the time of their lives tuning it all in.

This book will give you some good hints on what you can tune in, including some fairly rare and interesting stations such as tropical broadcasting stations, Coast Guard stations and distant broadcast stations which are normally hidden among the profusion of U.S. broadcasting stations.

In order to give you some of the fascinating history of radio communications we will introduce you to some of the early pioneers and their contributions, men such as Volta, for whom the Volt is named, Galvani (have you heard of the Galvanometer?), and probably the greatest electrical genius the world has ever seen, though he is little known today . . . the inventor of alternating current (our common house current), the ac generator, the ac motor, the transformer, the radio loudspeaker, the electric clock, fluorescent lamps, a man who was using radio communications many years before Marconi came along, who discovered X-rays, described radium years before Curie, and on and on and on. Who was this incredible man? Read on, my friends, as you enter the world of Radio Communications.

Wayne Green, Editor

Contents

How the Bug Bit Me

D. J. Holford

I guess I was about twelve when my father came home with the new radio. It was just like the old one except that it had an extra band marked "short wave" (how those words were to affect me) in addition to the regular broadcast band. As soon as the opportunity presented itself I began to explore this new band, to the accompaniment of strange and peculiar noises. However I soon located a human voice, and in the process discovered that this short wave business required a great deal more care in tuning than the regular broadcast band. The voice eventually identified itself as that of an announcer in France, and I was hooked!

My parents did not share my enthusiasm for the strange world of the short waves, and since this was before television reared its ugly head, I was only able to use the radio for my explorations on weekends or in the small hours of the morning. This did not increase my popularity with the rest of the family, as the noises found on the short waves at night have a particularly penetrating quality. I was saved by the foresight of the manufacturer in providing a connection for a headset. I recalled having seen an old pair of headphones in the science lab at school, and since the instructor seemed to think that I had possibilities of some sort it was a simple matter to obtain the loan of them for an indefinite period. Now I was able to sit up all night without disturbing the rest of the family. I soon discovered that not only were there foreign broadcasting stations to be heard; but there were also ships, aircraft, time signals, weather forecasts and amateurs. The short waves were truly a fascinating world full of new discoveries.

I have never been able to decide if my father permitted me to continue with this nonsense because he thought I might learn something, or if he realized that it was too late to save me, and that the only treatment was ever increasing doses.

Needless to say I began to acquire a collection of radio books and magazines, and learned that much more could be received with a better receiver and antenna. I constructed a simple long-wire antenna, and began to drool over the war surplus advertisements. I finally saved enough to order a surplus aircraft receiver. After an interminable wait it finally arrived, and I was shocked to discover that it had apparently been used as a football by a team using steel toed boots and then abandoned in a mud hole. But the advertisement assured prospective purchasers that all receivers had been air tested, although I began to wonder if that meant that they had been thrown through the air and if they landed in one piece they were considered to have passed the test. I cleaned out the first few layers of garbage – all the parts appeared to be there and all the knobs turned without too much effort – so I decided to give it a try. On inspection I discovered to my dismay that a separate power supply was required, and since my funds were exhausted I turned to my store of magazines and selected a simple power supply circuit.

The local radio repair shop supplied some old condensers and a damaged power transformer to use as a choke. The power transformer was another problem, but a canvass of neighbors and relatives unearthed an old transformer operated receiver which the owner was more than willing to donate to my project. The supply was hastily constructed and all was ready for the great moment.

I haywired the supply to the receiver, and harboring dark suspicions about the honesty of the surplus dealer, plugged it in. The tubes all began to light up, and after a few moments a faint hissing arrived in the headphones. IT WORKED! Signals came pouring in. A quick mental apology to the surplus dealer, and I began to enjoy my newfound bounty. I was enthralled and began to work my way from one end of the coverage to the other (70 kHz to 22 MHz). About halfway through my electronic journey I began to smell that distinctive odor of overheated components, a smell with which I have since become intimately acquainted. The power transformer was a small unit designed to power a five tube broadcast receiver, and was covered in a black pitch compound. This surplus monster was too much for it, and the pitch was bubbling away gaily, and dribbling down one side of the chassis. I rapidly unplugged the power and waited anxiously for it to cool down.

But the bug will not wait for funds to be raised for new transformers, so some experimentation revealed that although the transformer objected strenuously, it would operate without any apparent damage, at least for periods of up to four hours. I continued to operate with a bubbling power supply, much to the distress of my mother.

One evening after listening for some time, I turned off the receiver and went to a movie. When I returned home I found my receiver resting in a barrel of water outside the back door. My parents explained that shortly after I had left it had burst into flames, and closer examination confirmed that it had suffered irreparable damage. I never discovered what had caused the conflagration, but eventually managed to dispose of it to a fellow victim who was looking for parts.

It was many long empty months until I again located a receiver in my price range. This time it was an old RCA-AR-77 which had its own power supply, and went all the way up to 30 MHz. The AR-77 served with honor for a number of years, and introduced me – somewhat unsteadily – into the mysteries of single sideband, until it, too, met a violent end when it was dropped face down while being shipped to Canada.

I managed with a small National SW-54 for the next two years, as I was moving around a great deal, then progressed to an ancient Super-Pro, and at last the day arrived when I was the possessor of a 14 tube receiver of my own design which still rules the roost.

Although times and receivers have changed, as has the character of the short waves, there is still the same indefinable thrill at receiving some exotic or unusual station; be it a ham on a small speck in the South Pacific, a weak foreign broadcast station in some far distant nation, or even Air Force One carrying the President of the United States to some important conference across the ocean.

Broadcast Band DXing

Paul Schuett WA6CPP

You have tuned across the broadcast band and heard stations coming in from thousands of miles away. From the midwest you can hear the East Coast and Denver. From California you can hear Texas and Florida. This starts you as a broadcast band DXer – listening for the far sound between 535 and 1605.

Many of the stations send QSL (verification) cards to confirm a correct report; many do not. At any rate, you want confirmations from these stations. What's the best way?

For a number of years I was station engineer at several operations and received numerous reports from DXers around the country. Although my station did not have nicely printed QSL cards, I did send nice letters to the people who sent in reception reports. Some engineers put DXers down as pests and ignore the correspondence.

Let's have some positive suggestions for stimulating responses from radio stations.

Address your report to a specific individual by name at the station. This will assure the person probably will get it, instead of the mail girl who probably won't know what to do with it. Get a copy of the *Broadcasting Yearbook* (Broadcasting Magazine, 1735 DeSales Street, Washington, DC) which contains all sorts of information not immediately available elsewhere. The name of the chief engineer is given, along with the program director, manager, frequency, power, directional antenna information, alphabetical and frequency list of stations, and mountains of other data.

Generally, the chief engineer gets the reception report. If, after a reasonable time, you get no reply, send the report to the program director. As a last resort, try the station manager. If you still find no report, forget it; that station doesn't QSL. Remember that the personnel turnover is quite high; therefore it is best to avail yourself of a new Yearbook every couple of years or so.

Your reception report should contain the essential information in reasonably concise form. Probably the man at the station doesn't have time to read a lengthy letter. Your letter should be NEATLY composed on acceptable stationery. For a small fee, a local printer can run off a supply of forms for you to fill in – I was quoted $2 per 250 if I typed the offset master.

Be sure to have the date and time given in the local time of the transmitting station. If you are in New York and monitor Hawaii, give the information in Hawaiian time so that the station will not have to think of the conversion factor in adding or subtracting hours. If midnight intervenes, remember to change the date as well.

A few words about the program content – newscasts, times commercials heard and who advertisers were, name of disc jockey, etc., all help to verify your report. It is useless to give a list of the record heard since most stations don't keep lists of records they play – they operate on the "grab and spin" basis. One person sent me a tape he made of the station – but be sure to use the 7½ speed and wipe the tape with a bulk eraser before recording so the engineer can listen on the station's commercial equipment.

A brief word about your location, type of receiver and antenna, and any other pertinent remarks that you think will be helpful, should be included. Be sure to ask for a verification if you want one, and enclose a self-addressed stamped envelope.

There are several reasons for the envelope: First, it makes it convenient for the man to answer you in a hurry. Second, stamps and envelopes are hard to find in some stations that operate on low budgets. Third, the psychological effect works for you in that he won't want to waste the stamp.

Broadcast stations in the United States are divided into four classes and occupy three classes of channels. Clear channels have class I and II stations; regional channels class III stations; local channels class IV stations. The first three classes have subdivisions, such as I-A or I-B, depending on whether the station is a dominant station or has a secondary classification.

Generally speaking, you will have no problem receiving stations on clear channels, especially the dominant stations. Most of these stations will gladly verify your report; many have QSL cards for the occasion. The secondary stations on the clear channels can be received, often with some difficulty because of directional transmitting antennas and co-channel interference. The class II stations offer some challenges to the DXer, since they often operate with low power and directional antennas.

Work on receiving the regional channels. You will be able to hear these several states away; but you should try for the ones across the country, since these will be difficult. Most of these operate with directional antennas; none have more than 5,000 watts. Many operate directional antennas and/or reduced power nighttime; some operate daytime only, as do some secondary clear channel stations.

The local channels will be your biggest headache. Each of these channels is crowded with low-power stations operating 1,000 watts daytime, 250 watts nighttime and a few operating 500/100. Since there is excessive co-channel interference on these frequencies, you will need to do careful tuning and listening. It is not impossible – once I clearly heard a local station near Portland, Oregon, while I was in Blythe, California. While chief engineer at a local-channel station, I received a DX report from Nome, Alaska.

Don't be afraid to write to a station you are trying to log and ask for details on the technical setup. Most engineers like to brag a bit and will supply you with power, schedule, a copy of the coverage map showing the directional antenna pattern, and about anything else you want to know. If you live in a direction from the antenna in which there is minimum radiation, you will have difficulty copying anything; if you are in a direction where considerable power is radiated, you have a chance.

Some stations operate after midnight for frequency measurement purposes. The engineer can tell you when the next schedule is, and you can try to receive them on that schedule. Other stations use an off-the-air service and don't have the special program. Many operate periodically on "maintenance night" for testing and adjustment. When co- and adjacent-channel stations are off the air, you have a better chance of receiving the station.

About 10 years ago I was delighted to fire the place up about 2:00 A.M. so some guys 2,000 miles away could try to listen. If you find someone who is this nice to you, be sure to alert your friends so that several may catch the transmission. Also be sure to advise the man of the results, even if you didn't hear. Remember that people appreciate a thank you for services ABCD – above and beyond the call of duty.

KFI, Los Anglees, has a HUGE map in the back room showing locations from which reception reports were sent. There are pins all over the country – thousands of pins. I took this idea and have a map (not so huge) with pins showing locations I have contacted on my amateur radio. You may like the idea also to show locations of stations you have logged – using a different color pin for local, regional, and clear channels, or the various subdivisions.

UNITED STATES CLEAR CHANNELS

640	690	750	820	870
650	700	760	830	880
660	710	770	840	890
670	720	780	850	900
680	740	810	860	940
990	1040	1090	1140	1200
1000	1050	1100	1160	1210
1010	1060	1110	1170	1220
1020	1070	1120	1180	1500
1030	1080	1130	1190	1510
1510	1560	UNITED STATES LOCAL CHANNELS		
1520	1580	1230	1400	
1530		1240	1450	
1540		1340	1490	
1550				

The remaining channels are regional, Canadian Clear, Mexican Clear, Cuban Clear, etc.

Rand McNally supplied this on a framed cork board – it really adds to my living room. Even my wife likes it.

Go after the DX. Be aggressive in searching the local and regional channels. Don't be afraid to ask the station you are seeking for information, and when they might be operating after midnight for testing. Soon you'll log all states and can start in on the countries.

Tuning the Tropical DX Bands

Harry L. Helms, Jr.

If you have recently started as an SWL, you probably have logged and verified stations such as the BBC, Radio Moscow, Deutsche Welle, and other megawatt monsters. These are not, however, true DX. If you feel that you are to leave the multi-kilowatt and rhombic antenna equipped stations for one to twenty-five kilowatt powered stations with one element verticals, tune your receiver down a few megahertz to the tropical bands. Not only will you have to contend with the above, you will also have to fight through near overwhelming QRN, have to translate an ID that is in a language that is gibberish to you, and may have to send repeated reports for several years before a QSL arrives. But when that QSL arrives, you will have verified a piece of truly rare DX.

There are three tropical DX bands: 120 meters (2200–2400 kHz), 90 meters (3200–3400 kHz), and 60 meters (4750–5060 kHz). 120 meters is a place where only the true DXpert dares tread because of the low frequency, low power, and high QRN, a combination which makes any station heard on the band DX. 90 meters is not quite as difficult, and patience is all that is required to bag good DX. This band has the advantage of a frequency marker on 3330 kHz, compliments of CHU, Ottawa, Canada.

The best band, however, is 60 meters. An SWL with average equipment can easily log twenty-five countries in a few months of casual listening. You can log all continents, even two states!

Following is a continent-by-continent breakdown of what to listen for. These generally have some English and are good verifiers.

North America – Although not authorized to permit regular broadcasting on any of the tropical bands, the U.S. and Canada both have a representative in these bands. WWV, Fort Collins, Colorado, puts a potent signal on 5000 kHz each evening and makes a good frequency marker. It is also a good verifier. Canada, however, has a standard frequency and time station on 3330 kHz, and it is the only way to log and verify North America despite only three kilowatts. It gives an ID and the time every minute in both English and French. Like all stations listed here, it is a good verifier.

Central America – Radio Belize, in the tiny country of British Honduras, makes itself known on 3300 kHz with its usually good signals. Better still, about 50% of its programs are in English. It QSL's with a large card bearing a map of the country.

The "other" Honduras has a fine representative in HRVC, 4820 kHz. It is an evangelical station, and has some English religious programs and frequency English IDs. You may encounter heavy radio-Teletype QRM at times, so be prepared to wait for a positive ID.

South America – Station HJGG, Radio Sutatenza, is an exception to the rule that most tropicals use low power. It puts a solid signal into North America on 5095 kHz using a 50 kW transmitter. It is an educational station of the Catholic Church and has headquarters in Bogota, Columbia. Although sometimes slow, it verifies all correct reports with a bilingual folder.

Venezuela is represented by one of the most popular stations in the band. It is YVLK, Radio Rumbos, in Caracas. It operates on 4970 kHz and, although entirely in Spanish, has an ID that is easily picked out. It also is a good verifier, and accepts reports written in English.

Africa – One country you may have already logged is South Africa. Most likely this was Radio RSA and its 100 kW Bloemendal transmitter. But have you logged its internal service? The commercial service of Radio South Africa is noted well on 4945 kHz shortly before signoff around 0510 GMT. The transmitter is a twenty kilowatter located in Paradys.

If you really want to test your skill, try for Radio Clube de Mocambique on 4855 kHz. More often than not, this station will not be audible. It took me several months of almost nightly listening before I could obtain enough information to positively identify the station. But it is truly rare on this frequency and worth the effort. Try for it in the late evening hours.

Asia – Various regional outlets in Red China are noticed on the West Coast. Asia may be the home of the world's rarest DX station. It may be Radio Nepal, 4600 kHz. It is doubtful that there are a dozen QSLs from this station in the Western Hemisphere!

Also there are a few regional outlets of Radio Republic Indonesia on 60 meters. The best heard is on 4753 kHz. Right in step with today's headlines, you can tune in Radio Hanoi on 4713 kHz. They have English news at 1320 GMT, and you will probably enjoy their humorous rantings about "Yankee Imperalists," "valiant freedom fighters of the NLF," and similar drivel.

Oceania – If you have not logged Hawaii, you can do it on 5000 kHz via another National Bureau of Standards station. It is WWVH, and has a format like that of WWV except for the ID. How can you log it in North America with WWV on the channel? Well, you might have to burn a good bit of the midnight oil, but in the early morning hours (your local time) WWVH is heard quite well during the silent period of WWV. And if a severe ionospheric disturbance is occurring, WWVH may override WWV! My location is only ninety miles from the Atlantic, yet I have observed this even when WWV was in Greenbelt, Maryland.

If someone were to take a poll of SWLs to determine which 60 meter station was the most frequently heard from Oceania, the winner would probably be VLT4, Port Moresby, Papua. Its ten kilowatt outlet on 4890 kHz is heard often around dawn with good signals. An IRC with your report will bring a swift airmail reply.

We have listed all continents here except for Europe. The only broadcasters using the tropicals there are a few regional outlets in the U.S.S.R. These broadcast only in

Russian, and it would be wise to get some experience on the easier stations before trying these.

We have tried to show a sample of the fascinating listening to be found on the tropical bands, but there is much, much more. Keep informed on latest band happenings by subscribing to an SWL club bulletin or paper. A standard frequency list such as the World Radio-TV Handbook is an absolute necessity. An English-Spanish dictionary and grammar guide will be worth their weight in gold when trying to verify the majority of Latin Americans.

The tropicals are not just waiting to be logged, they are daring you to do so! So accept their challenge and good luck for a huge DX harvest!

Tuning the Coast Guard Channels

Jack White

The U.S. Coast Guard maintains a number of important services for mariners plying both the inland and offshore waterways about our country; and certainly not the least of these aids are regularly scheduled weather reports transmitted from certain strategically located stations. The information broadcast consists of storm warnings, and during the winter ice hazards as well as general forecasts and data relating to the region served by the station. Since these reports are intended to advise vessels operating in the local area of the station, tne medium wave channel of 2670 kHz is used for this purpose. This frequency of course cannot be propagated over great distances during the daylight hours, but there is really no particular need to do so as seamen interested in local weather conditions would normally be within range of at least one transmitter. However, as night falls, and the signal absorbing D and E layers of the ionosphere disperse, it becomes possible for these transmissions to be conducted via "sky-hop" over considerably extended distances – which brings the DXer into the picture.

Would you believe it possible to log fourteen states plus Alaska, Hawaii, Puerto Rico, and a ship at sea on one single uncluttered frequency? Take a look at the accompanying chart which lists the scheduled Coast Guard weather broadcasts on 2670 kHz through the North American darkness hours. Additionally, to put some frosting on the cake, sometimes during the silent periods unscheduled priority traffic between other of the Coast Guard's several hundred or more stations is also conducted. Such interesting sounding locales as Five Finger Light

Station (NMJ8), Galloo Island (NMD48), or Frying Pan Shoals (NNBM) located anywhere from Newfoundland to Alaska might be heard.

Before leaping to the receiver, the DXer is reminded that there are some factors that provide a measure of

Table I.
Schedule of U.S. Coast Guard Weather Broadcasts on 2670 kilohertz
(0000 through 1400 GMT)

CALL	STATION LOCATION	BROADCAST TIME (GMT)
NMD15	Marblehead MA	10 minutes past each hour
NMY	New York City	0020 1220
NUZY	Campeche Patrol Vessel*	0020 0620 1220
NMR	San Juan PR	0030
NMC	San Francisco CA	0200 1400
NMB	Charleston SC	0420
NOF	St. Petersburg FL	0420
NMF	Boston MA	0440 1040
NMA	Miami FL	0450
NMQ	Long Beach CA	0500
NMJ19	Ocean Cape AK	0510 0810
NMN	Portsmouth VA	0520
NOY	Galveston TX	0520 1120
NMW	Westport WA	0530
NOW	Port Angeles WA	0545
NMB	New Orleans LA	0550 1150
NMJ	Ketchican AK	0600 1430
NMV	Jacksonville Beach FL	0620
NMO	Honolulu HI	0630
NOJ	Kodiak AK	0700
NMD20	Belle Isle MI	0930
NMK	Cape May NJ	1100
NMN37	Fort Macon NC (Morehead City)	1130
NMD22	Port Huron MI	35 minutes past the *even* hour
NOG17	Portage MI	(0235, 0435, etc.)
NMP15	Plum Island WI	
NMD47	Buffalo NY	
NMD34	Charlevoix MI	35 minutes past the *odd* hour
NOG14	Duluth MN	(0135, 0435, etc.)
NMP12	Two Rivers WI	
NMD11	Erie PA	
NOG5	Marquette MI	55 minutes past the *even* hour
NMD23	Harbor Beach MI	(0255, 0455, etc.)
NMD41	Ludington MI	
NMD4	Oswego NY	55 minutes past the *odd* hour
NMD29	Tawas MI	(0155, 0355, etc.)

*The Campeche Patrol Vessel is located at sea in the Gulf of Mexico operating in the vicinity of the Yucatan Peninsula.

challenge to the matter. First off, this is medium wave territory and the going therefore somewhat rough for long distances except during the mid-autumn to early spring season. Next, these stations are not powerhouses by commercial station standards (usually only 2 or 3 kilowatts), so fairly stable ionospheric conditions are required to complete a transcontinental circuit. Lastly, the intended "audience" for these transmissions is located generally outward from the territorial perimeter rather than within it, so some broadcasts may be beamed away from a specific location, although most Coast Guard radio stations do usually employ non-directive antennas.

Requests for a QSL should be directed to the Officer in Charge of the particular radio facility and accompanied with return postage. Often the reply will be enclosed in postage-free envelopes in which case the stamps are returned, but sometimes it is an individual operator who answers at his own expense. In preparing a reception report keep in mind that Section 605 of the Communications Act prohibits divulging information derived from communications not intended for the general public. Weather broadcasts are considered to be intended for the general public and may be reported in detail, but any other traffic is strictly privileged with only the call and location of the station along with the time heard being allowable to mention. The Coast Guard is reputed as an excellent verifier and usually includes sufficient information to qualify a QSL for anybody's award program.

True, this is not classified as long-haul DX, but there are just enough problems involved to keep things interesting and possibly provide some diversion for a cold winter night. Highly sophisticated receivers or code proficiency are not required as all these broadcasts are in the AM mode and the channel is well separated from interference. If possible though, a long-wire antenna should be employed for best results.

Now that you have it, who's going to be the first in his or her neighborhood to catch an iceberg alert this winter?

DXing the Radio Amateurs

Harry L. Helms Jr.

Where do you find the most fascinating listening? Although some claim it's the utilities, others the police bands, the answer probably is the amateur radio bands. The variety of topics discussed over the air make for interest themselves. Also you will find that it is easier to log certain states and countries on the amateur bands than on the broadcast and shortwave bands. Add it all up and you can see why DXing the amateur bands is so rewarding.

In the United States, radio amateurs are licensed by the Federal Communications Commission. They are granted these licenses by the FCC only after passing examinations in Morse code (referred to as "CW" by amateurs) and a test in radio theory and radio law. The privileges granted depend on the level of difficulty of the exam. All but the lowest class of license, Novice, are permitted to run 1000 watts of power in their transmitters. One class, the Technician, must restrict operation to frequencies above 50 MHz. All others are permitted to use 1000 watts on the high frequency bands below 30 MHz.

Amateurs are permitted to use many types of transmission: CW, AM, radio-teletype (RTTY), and single sideband (SSB). We will concern ourselves only with SSB here, as this is now the most common form of transmission.

SSB is a form of radio transmission that gives greater effective power. This is accomplished by transmitting only one sideband. The carrier and other sideband is suppressed and all transmitter power is used to transmit the single sideband. However, the SSB signal cannot be received in the normal manner of an AM signal.

Your receiver must have a BFO or a switch marked "SSB" in order to receive SSB. Turn on your receiver and allow it to warm up. This is essential because a stable receiver is necessary to copy SSB. Turn on your BFO or SSB switch and tune it until the SSB signal dissolves into readable speech. You should be able to recognize an SSB signal by its gibberish sound, somewhat akin to a drunken Donald Duck.

There are various bands for amateurs, just as on the international broadcasting bands. There are divisions within the bands for CW and phone transmissions. We'll now discuss these bands.

80 Meters 3.5–4.0 MHz – The phone section runs from 3.8 to 4.0. This band is good for a few hundred miles during daytime. Night ups the range to over 1000 miles. Most foreign stations heard here are in the Americas. This band is extremely crowded at night, with a great many nets meeting here. There is a great deal of random conversation carried on (this is called "ragchewing").

40 Meters 7.0–7.3 MHz – Phone section, 7.2–7.3. This band is useful for short to medium range during daytime. Night range is several thousands of miles. This band suffers from severe interference from international broadcasters.

20 Meters 14.0–14.35 MHz – Phone section is 14.20–14.35. This band is good almost any time of day or night for worldwide reception. Many rare countries can be heard here. During good conditions this band seems to explode with DX.

15 Meters 21.0–21.45 MHz – Phone section is 21.25–21.45. Best during years of high sunspot activity and during daytime. When conditions are very good, this band tops 20 meters.

10 Meters 28.0–29.70 MHz – This band is useful during daytime and years of high sunspot activity.

Although you may hear some foreign languages (notably Spanish from Latin Americans), you will be amazed at the amount of English you hear. More than four-fifths of all amateurs live in English-speaking countries. In addition, virtually all books and magazines about amateur radio are printed in English. Since many amateurs in foreign countries are of upper-class backgrounds and are therefore highly educated, they have a good knowledge of the English language. Whatever the reason, the result is the same – an easy time for the listener.

It is quite easy to obtain QSLs to verify your reception from amateur stations. You should not repeat any of the conversations you might hear. Instead, you should give the call letters of the stations that he was in contact with, and the starting and stopping times of each transmission. You should express the time in GMT. Also, be sure to enclose return postage either in mint stamps or International Reply Coupons.

To find out the QTH (address) of an amateur, you should consult the *Radio Amateur Callbook*, which contains the names and addresses of amateurs along with their calls. It is issued in two editions, U.S. and foreign. The U.S. sells for $8.95 and the foreign for $6.95. It is available at most radio shops, or look for their ads in amateur radio magazines.

If you would like to keep informed about the latest in amateur DX, consult the International Shortwave League. They publish *Monitor*, which is devoted to amateur DX. Information on ISWL can be obtained from Bernard Brown, 60 White Street, Derby, England.

Try DXing on the amateur radio bands. You might like it so much that you could decide to become one!

The World of Medium Wave DX

Chuck Sadoian

If you glance at today's DXers, you might come to the conclusion that shortwave DXing is the "in" thing. Not so! While it is true that shortwave DXing is here more and more people are returning to the one and only medium-wave bands for their DX hunting.

Medium-wave DXing has been with us for over 40 years. In the beginning days of the broadcast band radio stations, there was born the medium-wave DXer. In those days, most of the catches he made were of 10 or 50 watt stations. Rare was the station with over 1000 watts of power. Tuning through the bands you would find stations few and far between. And as new stations came on the air to meet the demands of the public, the frequency squeeze was on.

It was not too long ago that Australian and New Zealand stations were semi-regulars on the East Coast, and England and the rest of Europe were semi-regulars on the West Coast. Today, it is a different picture. Powerful stations in all of North America crowd on the international clear channels, with the result being the near jam-up on the BCB channels.

Today's medium-wave DXer has to dig through this mess in order to get those rare and hard-to-hear stations. And when a DXer receives a verification card from a rare station, it is the Fourth of July. Unlike the SW bands, medium-wave propagation of signals is still a near mystery to most people. Although some research has been done in the area, there is still little known about how and why a medium-wave signal skips thousands of miles into a certain area at a certain time. The main reason that not much research has been done is that nobody but DXers care to know about it. Owners of BCB stations couldn't care less about their signals skipping from North America into Australia one day out of the year. In short, the stations don't see any profit in it. So, the DXer is left all alone in a world of strange, new signals, very much different than SW signals. If the DXer is fresh from the land of SW, he will find medium-wave DXing vastly different, and probably more exciting and rewarding.

In SW DXing, only on rare occasions do you receive a real "rare" station, because the bands are already crowded with the super-powered voices of the U.S., Russia, England, and China. But on the broadcast band, reception can change at any time you might be listening to your radio. From my West Coast position in the U.S., I would probably collapse from excitement if I picked up West Germany on the BCB. However, freak reception on the BCB is possible, even if you have a local radio station on the same frequency.

The best time of year to DX the medium-wave band is the winter months, beginning around October and ending around April. This is usually the "quiet" time of the year – that is, static and atmospheric noise is at a minimum. However, during the static-ridden summer months, many rare stations can be heard that usually cannot be heard. On the East Coast, the lower part of Africa comes through, while on the West Coast Australia is generally good.

The ionosphere also plays a part in medium-wave DX. When a "solar flare" shoots out on the sun, the usual result on the earth is what people call "auroral conditions." When solar particles reach the earth, they are drawn toward the earth's magnetic poles. When these particles hit the ionosphere, the gases are ionized so strongly that they absorb more of the medium-wave signals. The result is very interesting. If there is a large enough solar storm, the "auroral storm" can extend clear over the poles, plus half-way to the equator. When you tune the band on a night like this, you will notice an increase in noise. Then you will begin to notice that a lot of your surrounding stations are fading out, and stations from the South, such as Central and South America, come in with terrific strength. This is caused by the auroral. The "blanket" extends over such a wide area that the stations that usually have a clear path are knocked out of the sky.

This is only a sample of what medium-wave DXing can be like. It can be very exciting, and it can be educational, too.

DXing Television and FM Radio

Henry L. Helms, Jr.

When the FCC assigned channels for television and the FM broadcast band, the choice was made to put them in the VHF and UHF ranges because it was thought that nothing could affect reception at such high frequencies. However, it was later discovered that an ionospheric phenomenon known as Sporadic-E could cause abnormal reception, as well as certain weather and astronomical factors. Although this may be a pain in the neck to the average viewer, it affords the DXer with an opportunity to make some nice catches.

Sporadic-E is the main cause of abnormal reception. It usually affects television channels 2, 3, 4 and 5 as well as the FM broadcast band. Although not fully understood, Sporadic-E occurs in the ionosphere at an altitude of about 75 miles. Sporadic-E is believed to take the form of a cloud as it travels. It reflects those VHF and UHF signals that would usually pass through the ionosphere and out into space. Since the cloud appears to move, signals may last for hours or minutes.

Another method of receiving DX signals is tropo reception. This usually occurs in spring when air masses conflict to form a reflecting mirror similar to the Sporadic-E cloud. The conditions which lead to tropo reception are, unfortunately, similar to those which form tornadoes.

Recently research has begun into meteor scatter reception of FM signals. This most frequently occurs during meteor showers, and the signals have a short, hissing quality. Sporadic-E and tropo are noted by rolling black bars on the lower TV channels.

You should use an outside antenna for TV and FM DXing. However, do not use an external preamp. It will increase interference as well as the desired signal.

TV DX is not restricted just to U.S. stations. I have also seen Canada, Cuba, and Mexico during the space of a year.

It is often difficult to get verifications from TV and FM stations. Quite often, the station will not have the slightest idea of what you want. It is always wise to include a prepared card, although some stations do send nice cards or letters.

If possible, try to photograph the station's identification card off the TV screen. Have two prints made, send one to the station and keep one for yourself. This will aid in getting veries and will prove your reception in case the station doesn't answer your report.

If you get "hooked" on TV and FM FXing, we suggest that you write to the Worldwide TV-FM DX Association, Box 5001, Harbor Station, Milwaukee, Wisconsin 53204. They publish *VHF/UHF Digest*, the bible of TV and FM DXers.

TV and FM DXing is like BCB DXing thirty years ago – just getting started. This is your chance to get in on the ground floor of a potentially big hobby.

Amateur Radio—What It Is

Jim Barcz WA9JMY

Have you ever heard anyone mention amateur radio and wondered what they were talking about?

Amateur radio is a government licensed hobby with about 350,000 members. Anyone regardless of age, sex, or creed can obtain a license. In order to obtain a license, it is required that the applicant pass a written examination and a test of his ability to copy and send the Morse code. There are five different levels of licensing: Novice, Technician, General, Advanced, and Extra. A detailed explanation of the privileges and requirements of each license is beyond our space limitations here, but suffice it to say that as the privileges of the license increase so do the requirements. At any rate the tests are designed so that it is possible for anyone to pass, providing they study. There are fellows who have passed their Novice test at the age of 9. If a 9-year-old can do it, so can you.

Once possessing a valid license and call letters (the call letters are issued with the license) the amateur has an opportunity to operate on certain radio frequencies set aside for his use. Since amateur radio is made up of so many different types of people, it follows that the amateur frequencies will be used in a number of different ways. It is the combining of all these interests that make amateur radio an intensely absorbing hobby. I will now explain the different phases of amateur radio with the purpose of arousing the reader's interest to the point of obtaining a license for himself.

I will quickly mention here that a common expression for amateur radio is ham radio, and the amateur is called a ham. From now on I will use the term "ham."

Rag Chewing

By far the most popular use of the ham frequencies is for the purpose of rag chewing. Rag chewing is the art of yakking with a fellow ham via the radio waves. This fellow may be a close relation or a total stranger whom you have never met. The ham prides himself on the fact that he is considered a friendly Joe and will do anything in his power to help his fellow man. It may be surprising to the reader to find two people who have never met talking over ham radio like long-lost brothers. This is ham radio.

The following is a quote from a letter written by John Almber (WN2IEM) of Smithtown, New York to a popular ham magazine. I feel that John's words say it all:

> "Amateur radio. Just what is this strange hobby of ours? It's two raw novices fumbling with words, learning to speak with their fingers, finding the old Morse code is a lot of fun after all. It's a couple of two-letter men on the low end of the band talking of their grandchildren, and roses, keeping off old age. Or perhaps it's one of each, bridging the age-old generation gap."

Kind of restores your faith in mankind, doesn't it?

You might wonder what two hams talk about without knowing each other's interests. This is simple to answer. Most hams have a way of getting the other fellow to talk about his other hobbies. Many times there is something that the two have in common. Presto! You have a lifelong friend.

It is also possible for more than two hams to carry on a conversation. Sometimes a whole group of three to ten hams gather to form what is called a round table. Each ham gets his turn to talk while the others listen. Most of the ham's time in a round table is spent listening. You may think this is boring, but nine out of ten times listening and learning is more fun than talking.

There are also groups that get together to discuss a certain subject. For example there is a group that meets on a certain frequency on Wednesday nights to discuss UFO's. If this subject intrigues you, imagine what fun you could have listening to other fellows talk about it. The ham is known for his rag chewing ability. I have given you an idea of some of the ways a ham rag chews, but by no means have I pointed out all the aspects of rag chewing. Much is left up to the ham himself.

DXing

DX means distant stations. Since ham radio is a worldwide hobby, there is a drive inside the average ham to see how many different places around the world he can contact. The process of trying to contact different DX stations is called DXing. In some countries there is an abundance of ham operators but in others there may be only one or two. It is these countries that are the target and hopeful possession of the DXer.

Many DXers go to great measures to improve their station to the point where it will be easier for them to hear DX and for DX to hear them. Some do this by building efficient antenna systems. Others do this by running high power. I should point out here that the average station with a transmitter, receiver, and a modest antenna has the capability of worldwide communication. The purpose of putting up hugh antennas and running high power is to make one's signal superior to the next fellow's signal. By doing this he will have a better chance of contacting the rarer DX stations. I want to emphasize again the fact that anyone with a modest station is capable of working many DX stations.

There are incentives for the DX operator to shoot for in his hunt for as many countries as possible. These incentives are certificates. Although certificates are just worthless paper, they are the treasured possession of a ham. Certificates boast of his achievements in certain areas. They also attract the attention of a non-ham friend who is visiting and serve as a convenient conversation piece.

The first certificate that a DXer can obtain is called the DXCC. He obtains this when he can prove that he has contacted 100 or more foreign countries. Once this certificate is obtained, the ham has an opportunity to receive endorsements to place on his original certificate.

These endorsements are given out to holders of the DXCC when they can prove they have worked more DX stations. The endorsements range from 120 up. The highest endorsement I know of that has been obtained is 340. There are not too many hams who can achieve this many, but it is something to shoot for.

In any hobby there is always the factor of prestige. In ham radio, the DXCC with as many endorsements as possible is a prestige builder.

I might mention that not all hams are interested in contacting DX for the purpose of contacting the most countries. Some have interests that lie in other facets of ham radio, and only make a casual DX contact now and then.

Public Service

One of the main reasons the government favors ham radio is because of its service to the public. In times of local, state or national disaster, the amateur makes his equipment and operating skills available for public use. He is always willing to help his fellow man whenever possible. Remember Hurricane Camille? This fantastic storm created so much destruction it is almost incomprehensible. Do you remember the hundreds of small towns cut off from the rest of the world by Camille's force? Had there not been countless hams who devoted their time and equipment so that communication could be restored, the Red Cross and other disaster agencies could not have located these people. Imagine if there were no such thing as ham radio. These people would have been trapped, with no telephone, telegraph, or any other form of communication. It could have been weeks before disaster teams found them, and by that time it might have been too late.

As you know, a major disaster such as Camille does not happen often. This does not mean that the hams' public service stops. Not by any means.

The amateur organization is arranged in a strategic way so that messages can be passed across the country efficiently. This system is called the NTS or the National Traffic System. This system is made up of various nets around the country. A net is a group of tightly disciplined hams getting together on a designated frequency at a designated time to handle messages. The purpose of these nets is to form a channel across the country through which messages may pass. Let us take an example. Suppose you were a mother in New York who had a son in Vietnam, and it was important that you get a message to him concerning something that happened at home. This is how you would do it: You would get in touch with a ham; he might be a friend or just someone you knew of. You would give your message to him and tell him where you wanted it sent. That is all you would have to do. The rest would be up to him and the NTS. The ham would pass the message along to a local net. This net in turn would send the message to a net located further west. This procedure would continue until your message reached the authorities in Vietnam, who would locate your son. All this within 48 hours, and at no cost to you! Amateur radio does not charge for its services to the community.

Another way to handle traffic is by what is called phone patches. This is how a phone patch works: Suppose a ham from Newport, Rhode Island contacted a ham in Houston, Texas by pure chance. It so happens he has relatives in Houston with whom he would like to talk. He asks the fellow in Houston to run a phone patch for him. If the Houston ham has the proper equipment, he will usually agree. The Houston ham then places a phone call to the person that the Newport ham wants to get in touch with. When everyone is ready, the telephone is connected to the receiver and transmitter so that whatever the Newport ham says will be sent over the Houston telephone lines and whatever the Houston party says will be transmitted by radio waves to Newport. Thus two-way communication is achieved with no cost to the Newport ham. Phone patching is not limited to within the U.S. but may be conducted between different parts of the world, such as the U.S. and the Antarctic.

So you see, ham radio plays an important part in speeding messages across the country. You might say that amateur radio strengthens the chain of communication.

Moonbounce and Satellites

With the great attention focused on the moon by all Americans, the topic of moonbounce should prove interesting to many.

Moonbounce is the process by which a signal originating on earth is beamed to the moon, and on striking the moon it is beamed back toward earth. Although communication is not as dependable as by other means, it is possible to establish communication via the moon. There have been great efforts to establish better communications, and one of these efforts is the construction of high performance antenna arrays such as the parabolic type antenna. It is important to keep in mind that the antennas used in moonbounce activity are extremely important. Equipment is also very important in this type of work.

Satellite operation is another fascinating aspect of ham radio. A number of satellites have been constructed by amateurs and sent into orbit by the government. These satellites were designed to send transmissions back to earth.

At the time of this writing, OSCAR 5, the fifth ham satellite, is being prepared for launching. This satellite will be different from the others in that it will send back data that will be useful in future flights. This data will contain temperature information, condition of batteries aboard, and the position of the satellite in space. The nice part about this satellite, is that anyone can receive and interpret this data. It is not necessary to have a license to listen. Only a receiver and antenna are necessary. The data is sent back in the form of tones. A complete explanation of the interpretation of the tones appears in a popular ham magazine. Wouldn't it be fun to tell your friends that you track satellites for fun?

Experimenting

Some of you may like to experiment with building little gadgets. Wouldn't it be stimulating to build a transmitter and be able to try it out on the air and experiment with it? The door is open for the experimenter to exercise the imagination of his mind in ham radio. By holding a license, he is allowed to run tests on the air. The test may be on a special UHF transmitter to see how it functions under different operating conditions, or the test may be with lasers or masers. The experimenter may try to modulate a laser beam and thus maintain communication with a helper who is receiving the laser a few blocks away. All this may seem far-fetched, but as I said before, the door is open. The amateur bands extend into the Super High Frequencies and the Extremely High Frequencies. These bands are open for experimentation. In fact, the government encourages it, for they realize that many developments start in ham radio.

Homebrewing

Homebrewing is the art of building your own equipment. Back in the days when ham radio was young, all amateurs made their own equipment. They wound their own coils and made their own capacitors. To them, half the fun of amateur radio was in the building of equipment and watching it work. These men received a great deal of satisfaction from their hobby. But like everything else, this changed. Today there is a great abundance of commercial equipment on the market. Today it is possible for a ham to own a very sophisticated station without knowing how to solder a wire. This is sad, but it is not as bad as it sounds. Today there is a percentage (I would say about 20%) who build their own equipment. It is these hams who must take it upon themselves to encourage the newcomer to homebrew.

There are many excuses for not building your own gear and some are good, but they are only excuses, not reasons. I know a ham who obtained a DXCC certificate, a feat that takes a great deal of time for the average ham. I asked him why he didn't homebrew and he replied, "No time." My theory is that if one has time enough to operate, he has time enough to homebrew.

You may be wondering what the advantages of homebrewing are. I think the best way to hook you on homebrewing is to give you an example. Suppose you have just obtained your Novice license and you now want a rig so you can get in on the fun of the ham bands. You have two alternatives; either build your own or fork out the cash for a ready made job. Here is the difference – a new piece of commercial gear suitable for the Novice would run anywhere from $70 to $150. The same rig could probably be constructed for $10 to $15. It is possible to make it for less, if you can dig up a few used parts. In a recent issue of a certain ham magazine, the author presented a Novice transmitter that could be built completely from the parts of an old TV set. If you could dig up an old television set from a neighbor or friend (this is usually no great problem) you could build a working transmitter for no outlay of cash at all. I might mention that as equipment gets more sophisticated the manufacturers' prices soar. Let me give you another example. I am in the process of building a complete single sideband station. My present project is a six band transceiver. The estimated cost of this transceiver is about $200. When complete, the transceiver will be comparable in looks and operation to a rig running anywhere from $800 to $1000 on the commercial market. What's more, I will have the satisfaction of creating it with my own two hands. I have built before, and take it from me, it's a marvelous feeling of accomplishment!

Let me now say some final words about homebrewing. Homebrewing is a combination of emotions: joy, frustration, ecstasy, heartbreak, and confusion. One thing will render itself true in the end, though, and this is the fact that you will be congratulated a hundred times over by hams on a job well done.

Conclusion

I hope this little explanation of ham radio has aroused some interest among you. As you have well seen, ham radio is a very diversified hobby and one that has the ability to appeal to almost anyone. If you are interested in learning more about ham radio, I have a few suggestions. First, visit your nearby electronic supply store and see if you can pick up a few publications concerning ham radio. There are many good ones available. Second, visit your library and ask to be shown where the books concerning amateur radio are kept. Third, contact a local ham. I have mentioned before that the ham is always willing to do anything in his power to help. A local ham will be able to do things that books cannot always do. He can always answer questions. He will give you advice and may help you when you are building your first transmitter. A ham friend can definitely be a major asset in your struggle to obtain a license.

What to Look for in CB Gear

Robert M. Brown

Conservative market studies by leading citizens band equipment manufacturers, merchandisers, and trade publications conclude that the average CB'er is not satisfied with the equipment he now has. Further, like the young man with his first automobile, the longer he remains a CB'er the more discriminating his needs become. Within the first three years of operation, the typical CB'er runs through more than three transceivers, generally upgrading in the process. At any given time 73% of the CB population are displeased with operational limitations inherent in the equipment they're now using.

What's behind these statistics? Is there an underlying problem connected with the purchase of CB gear? Are there guidelines that should be followed in evaluating new and used transceivers?

As the discerning CB'er eventually learns for himself, no single transceiver will ever contain all the generally desired features within the price bracket the buyer has in mind. Further, many operational features which look good in sales literature and advertising often are of relatively little significance while some overlooked major function is what is really required. Both potential and experienced CB users often have differing equipment requirements in mind.

Which all tends to confuse the basic question: "What do I need?" Personal preference will always be a strong governing factor in determining features such as tunable vs fixed receive, optional or universal power supply, IC or standard transistor configuration, number of channels, etc. Tempering preference with knowledge, however, can save you a substantial amount of cash over the weeks and years ahead.

First Things First

Stripped of all the glamor, playing the CB selection game is, after all, not much different than buying a car. That sensational little foreign car can be a city dweller's dreamboat, yet in the farm country a nightmare. Why? For starters, try locating a service dealer who stocks the parts you need when she conks out. By the same token, your superselective CB walkie-talkie in a metropolitan area may be the difference between keeping contact and not – but in the boondocks you may find yourself wishing you'd invested instead in a rig with a more sensitive receiver to pick up those weak, distant stations.

And there will always be people anxious to save every penny by plunking $450 on the counter for a very old Detroit gas-eater content to admire themselves for all the depreciation losses they don't have to worry about . . . until the beast churns to a final halt sixteen blocks from the used car lot. Rather like the skinflint metropolitan CB'er determined not to spend over $40 for his transceiver, only to find when he's hooked up the rig under the dash that

For this Mack truck driver, facing big-city channel congestion plus extremely high ignition noise, utilitarian CB equipment is required – with an eye to effective noise suppression, high selectivity. Clearly expensive rf amplification stages and extreme signal sensitivity make no sense for this application.

For a country mobileer, base-station performance equipment might indeed be a good investment. Emphasis should be on sensitivity; selectivity not being critical due to sparse local CB population.

he's plagued with ignition noise and when a signal does appear it's on six channels at once.

The answer? Simply put first things first. Instead of asking yourself, "Which transceiver do I need?" try starting with, "What are my needs?"

In the marvelous, chrome-lit world of the CB marketplace nothing comes free. Bear in mind that you are going to pay through the nose – in cash – for every operational feature your rig contains. Whether you need these features or not. For example, you may find yourself with five times the speech clipping capability you'll ever require, yet for all of your, say, $225 investment, no mobile mounting bracket. Or perhaps the world's greatest "base or mobile CB transceiver" (according to the magazine ads) – yet lacking a universal power supply capable of immediate switchover from household power to mobile power.

Learn as much as you can about CB equipment features – then narrow your choices down to those only that have direct, immediate value to you.

Some of what follows may be familiar to you; a lot will be new and a bit foreign. Spend your time, for the most part, on acquainting yourself with some feature or circuit advantages you can use later in evaluating future CB rigs. There is no better road to getting your money's worth than understanding what it is you're buying.

Full Coverage?

Unlike ten years ago, today you can select from a host of transceivers, offering every conceivable combination of channels – from the one-channel monitor rig to the frequency-synthesis, full-coverage model. Even better (or worse, depending upon viewpoint), there are 23-channel transceivers with built-in shortwave, AM, and VHF/UHF receive switchover capability! Want more? How about the 23-channel base stations offering business band switchover? It's so easy to lose sight of the fact that few CB'ers every really *use* all 23 channels.

Assuming, for example, that you feel a need for full 23-channel receive, think about how many channels you really ever transmit on. Perhaps a 23-channel tunable rig that comes equipped with only one or two transmit crystals is the ticket for you. If you'll be confining operation to five channels 99% of the time, maybe it's worth the cost of a few extra crystals over-the-counter as needed, compared with the often considerable investment for frequency synthesizing networks that eliminate this need, but at a premium.

Full-coverage transceivers always provide this advantage at a significant price – whether you need it or not. If you are an exceptionally active CB'er, however, who participates in a number of community activity groups, REACT and the like, you might actually be saving money by buying 23-channel full coverage at the outset.

Exactly how many channels you need is your business, and yours only. No one can make this decision for you, unless you simply don't know any better. If you are a novice at the CB game, hold off making any significant purchases – borrow a rig, monitor for a while. Get to know your personal needs before letting loose with your hard-earned cash.

Transmitter Output Circuit and TVI

TVI – or television interference – will probably not be a concern if your primary application for your CB transceiver is as a mobile unit. If, however, you're looking into a base station unit, you might want to have a look at your prospective rig's transmitter output coupling circuit – where most TVI emanates.

Not as popular these days, but still found in many low-priced or vintage transceivers, is the coil-capacitor "tank" circuit. What happens here is that a movable coil is used to load the antenna to the amplifier tank circuit while actual amplifier tuning is accomplished with a tuning capacitor. Unfortunately, this coil link is difficult to adjust and one of two things can happen when jarring or vibration changes its position: Too much coupling overloads the final transistor or vacuum tube, or not enough coupling provides insufficient output power to the antenna.

More important, conventional link circuits provide no suppression of "harmonics" (transmitted signals at a multiple of the desired 27 MHz frequency) whatsoever – so the problem of reducing TVI becomes difficult.

More often than not, however, you'll want to investigate transceiver's using pi-network coupling configurations. These systems typically afford two controls for the technician to adjust: amplifier tuning and amplifier loading. Efficiency is generally good and the circuit's ability to suppress harmonics of the desired output frequency is even better.

Most CB transceivers – though not all, unfortunately – employ low-pass filters in their output circuits to attenuate radiation of interfering harmonic signals. These filters vary in performance, but generally are a lot better than none at all. Bear in mind that in addition to being illegal, output at twice your transmit frequency plunks you down quite well from channels 2 through 4 on nearby television receivers. The better of these built-in filters allows you to adjust by means of screwdriver for maximum reduction of interference in your particular installation. If you overlook this feature in purchasing a CB rig, count on laying out additional cash later on for an external coaxial low-pass filter.

Your Modulator

The Federal Communication Commission limits modulation to the 100% level – something all manufacturers watch closely. It does not, however, require CB sets sold in the U.S. to be capable of this much "talk power." The most common circuit used is what designers refer to as Heising; it's economical, but unfortunately results in distortion when forced to operate in excess of 80%. Revised Heising circuits, which include an additional capacitor and transformer, work effectively with minimum distortion at modulation levels up to 100%.

In most transceivers today, however, the problem is not one of achieving 100% modulation but rather one of keeping it from being overdriven by external devices attached by overzealous operators. Several methods of

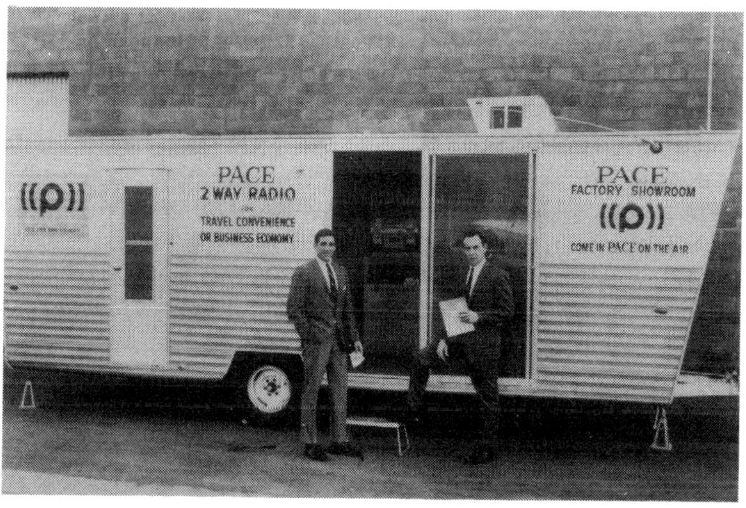

Several manufacturers have traveling mobile showrooms where CB'ers can actually operate, one by one, a company's entire line of products. An interested CB club can often be successful in talking a manufacturer into stopping in their city.

Even a quick browse through your corner electronics store will reveal that when it comes to selecting CB equipment, none of the so-called extras are free. In smaller communities, dealers are sometimes amenable to allowing CB equipment to be taken "on approval" for a few days prior to purchase. If an opportunity like this comes your way, grab it! Nothing is better than testing the set ahead of time in your own home environment.

modulation circuitry are offered in addition to that described above (variations on plate modulation, screen-grid modulation, etc.), but for the most part the only rigs to be leery of are the extremely inexpensive tube sets which have been known to provide inadequate delivery. It's interesting to note that while the market abounds with external goodies for boosting talk power, the rigs that need them the most are usually in the hands of penny-pinchers content with 50 or 60% modulation.

Range Boosters, Compressors, Clippers

Hardly a rig is offered today without at least one of the following: range boost, speech compression, or speech clipping. Contrary to popular belief, these features – which, of course, you pay for dearly – are of no significant advantage unless your signal level at the receiving end is extremely weak. To an S2–3 signal, for example, they can provide the extra "talk power" that makes the difference between understood and not. No advantage is provided in situations where the signal is received at S7–S9 levels or in interference-free circumstances. For long-haul communications, however, you'll find these features valuable.

Your Transmitter's Final

Output efficiency is the key to achieving optimum performance in today's class D CB equipment – regardless of whether it is of tubes or transistors. With input power to the "final" limited by law to five watts, clearly you want as much of this as possible to be transferred to the antenna.

In tube sets, avoid buying a set where the final amplifier tube is used as both the oscillator and rf power amplifier. Better that a separate tube be used exclusively for the final amplifier. Why? Better operational efficiency.

Be suspicious of any manufacturer who claims more than 3.5 watts rf output from the transmitter's final. If you still, however, can't resist trying the transceiver rated at "4 watts output," take the unit out on approval and have its output accurately measured with a Bird Wattmeter or other reliable instrument. With five watts input (the legal limit) it's pretty hard to get more than 3.5 watts out. That's 70% efficiency – and all you have a right to expect from today's rf amplifiers.

A few years ago – before Federal Trade Commission investigations into CB advertising – it was the vogue among some manufacturers to tout "10-watt construction" or even "25-watt circuit configuration" in their claims. This was paralleled by asking a price for the transceiver directly proportionate to the difference between the legal five-watt input and the wattage rate claimed to be built into the transceiver. The message was clear: this particular transceiver could be "modified" by the user for 10- or 25-watt input operation. The manufacturer simply employed a plate voltage sufficient to drive the higher-powered circuit, but lowered it by installation of an appropriate series dropping resistor for over-the-counter sales; as actually furnished, measured power input would be in the order of 4.94 watts. After the smirking buyer got his set home, he'd check a notice in the instruction manual that read: "WARNING: DO NOT SHORT-CIRCUIT RESISTOR R4: DOING SO WILL CAUSE TRANSCEIVER TO OPERATE ILLEGALLY AT 25 WATTS INPUT." Guess what resistor the buyer would immediately jumper with a piece of bare wire . . .

Needless to say, the Government put a stop to this activity, although a few of these older sets are still to be found for sale. Steer clear of these sets; by current interpretation, even the CB'er caught merely in *possession* of such by an FCC representative has a lot of explaining to do.

In short, stick with known quantities. If you must consider exaggerated advertising claims, check the equipment out first with reliable wattmeters before you buy.

Selective Calling Devices

While mostly available as an optional accessory, a few Class D 27 MHz CB transceivers come equipped with built-in selective tone calling devices. Still others are designed to accept accessory attachments specifically designed for that brand's system. Remember that you are paying dearly for the feature. Is it something you really need?

If what you are looking for is indeed a private communications system – meaning two-way CB conversations *only* with your Unit 1 or base buddy – selective call is the only thing short of SSB (single-sideband) that can come close to guaranteeing exclusivity.

A selective call signal acts to mute the loudspeaker through the audio circuitry until it receives – through a decoding process – a predetermined specific audio tone or series of alternating pulsed tones. By the same token, selective call also acts to "preview" your transmission with an encoded signal that can only (in most cases) be decoded by the mating unit attached to or contained within the receiver of the station you are calling. Most selective call devices are capable of providing several codes – enabling one base station, for example, to "unlock" several individual mobile units or other bases operating on the same channel.

While still somewhat the state-of-the-art in private communications remember that your transmissions will still be overheard by other CB'ers. The only thing "private" about the system is that you won't hear them.

Switchover "Popping"

Nearly all CB transceivers are equipped with a push-to-talk switch on the microphone that couples to a relay in the transceiver that is responsible for receive-to-transmit switchover. A few transceivers – specifically base stations using desk microphones instead of hand-held mikes – instead use a lever switch or toggle switch to accomplish much the same function. Some, in fact, don't use a relay at all – as in the case of a few of the lever switch configurations – where all switchable circuitry is, in fact, wired to the manually operated switch.

The only thing to watch for – and you'll have to try out the transceiver in order to find out – is a distinct audio "popping" sound from the loudspeaker that occurs when switching from receive to transmit, and sometimes even vice versa. Usually, it's only a somewhat disturbing annoyance to the operator; it has been known, however, to damage the receiver "front end" as a result of split-second overload.

As an example of extra features, consider the front-panel offering above. Of what is present, only the on/off/volume, squelch, and pushbutton channel selectors could be considered standard equipment. The extras? Tunable receiver, the tune/xtal "spot" switch, the S/rf meter and switch – and particularly the meter's own on/off slide switch.

Similarly, ask for a report at the listening end to determine whether any mechanical popping or squeals are audible. They can be murder to listen to for any length of time. If you're looking at a lever-operated design, try very slowly activating the switch. If it is properly designed, the first thing that will happen is that the receiver will be disabled *then* (and only then) should the transmitter light come on, indicating actual rf applied to the final or antenna.

Your Crystal Socket

If you don't have frequency synthesis, you are going to, at one time or another, want to replace or add transmit and receive crystals to your rig for new or additional channels. While all transceivers have the crystals enclosed in the unit – access usually via cabinet removal or a trap door – a few units come with a crystal socket or two furnished on the front panel. While this usually signifies a limited-coverage transceiver, it can be a handy feature, saving you considerable time and aggravation. This is particularly true when the receiver is tunable and all that is required for new-channel operation is installation of a single transmit crystal. The concept – a carry-over from broadcast and amateur radio equipment manufacture – is worth looking for.

By the same token, however, it can be a decided disadvantage to the mobileer who is likely to cause inadvertent damage to an exposed crystal on the front panel. Aside from accidental physical jarring, direct sunlight can zap the most expensive CB crystal.

Modulation/RF Meters

Front panel meters supplied on contemporary CB transceivers for the most part do little more than indicate there is *some* modulation present. If you must have this consolation, at least watch the price tag. Those meters that are allegedly calibrated to show percent modulation are in a word a "disaster." Only in very expensive, top-of-the-line base station equipment can you expect to find anything even remotely relating to accuracy. For most people, a single bulb that lights on transmit and blinks somewhat in response to modulation is more than sufficient for purposes of reassuring them that the rig's operationally okay.

What's an rf meter? Let's confine our discussion to the power-output meters, either of the relative power or calibrated configuration. Like their modulation meter counterparts, the calibrated rf meters are consistently inaccurate and misleading. In most cases they are little more than relative power output meters calibrated with a power output scale that is accurate only in a resistive load situation – which most antenna installations are not.

Any type of field-strength or relative power meter, however, will be useful in tuneup. For regardless of the degree of power being shown it will indicate when a peak of rf has been transferred to the antenna. As a tuneup meter – and nothing more – this meter is well worth having. Provided you appreciate that the circuitry can be purchased separately in a neat package from most radio parts catalogs (as FSM) for $5.95 or less. Paying $30 extra for this feature would require some soul-searching.

SWR Meters

While the traditional field-strength meter does show output, it must be remembered that it is relative. What this means is simply that while you are indeed tuning for maximum reading you really never know if that reading reflects the *highest power* your CB system can produce. What an swr (for standing wave ratio) meter indicates is the overall efficiency of your transmission line and antenna.

Swr meters only recently have been available within CB transceivers. For most users, these devices still have to be purchased separately at a cost probably averaging around $15.

What the meter does is to sample the rf power being sent to the transmission line and antenna when the meter's switched to its *forward* position. As you flip the switch into its *reflected* position, however, the amount of power returned to the transmitter is measured. This is compared with the amount of forward power, giving you an indication of what percentage of your signal is going down the drain due to a mismatch somewhere in your system.

If your swr meter reads 3:1, it indicates that 25% of your signal is going back into the transmitter; a reading of 2:1 indicates an 11% bounce-back. An indication of 1.5:1 translates into 4%, which under most circumstances can be tolerated. Obviously, when a substantial amount of rf is reflected back to the transmitter you lose a lot of performance.

Far worse, however, you run the risk of blowing out the transmitter's final transistors or tube. It can readily be seen, then, that if a dangerous degree of reflected power can crop up as a result of an antenna mismatch, running – even for a moment – with no antenna at all for test purposes is committing suicide. Walkie-talkie users with telescoping antennas should also keep this in mind; unless the antenna is extended to its full length, so much power will return to the final stage that the transistor circuitry may indeed blow itself out.

The Dummy Load

Never hit that transmit button without first giving your transceiver a load equivalent to that of an efficient CB

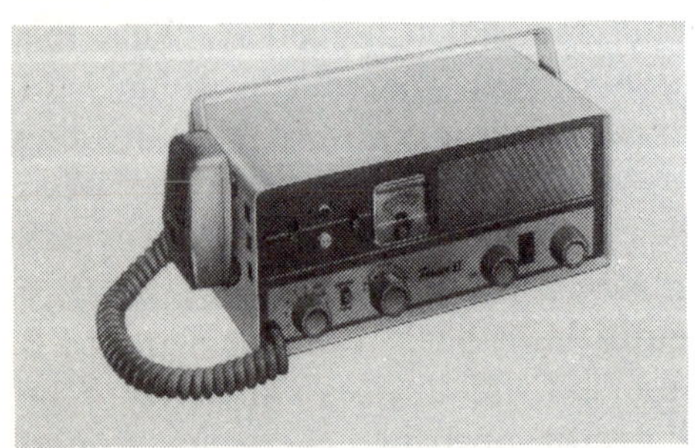

In the top photo is a typical example of a tube-type, continuous tune CB transceiver. This is contrasted with a typical eight-channel fixed-tuned solid-state entry from the same manufacturer.

antenna system. Several of the more expensive CB transceivers feature a built-in dummy load that can be switched into the circuit for test purposes. For others, a dummy load must be purchased or constructed.

If evaluating several CB transceivers – all featuring "built-in dummy load" – see if you can determine whether any one of these is shielded. A shield, which can be nothing more than a tiny can grounded to the chassis, provides insurance that when you are testing your rig the whole world won't be eavesdropping. Another good reason to look for this feature is that it is illegal in the eyes of the FCC to test on the air.

Tube or Transistor?

Let's list a few pros and cons here and let the jury decide, based on individual requirements.

Transistor advantages: 1)instantaneous warmup, 2)low-profile, space saving construction, 3)relatively low cost, 4)low power drain.

Tube advantages: 1)generally hand-wired construction, easy to service; 2)tubes are replaceable; 3)tuning controls are generally accessible.

Transistor disadvantages: 1)tuning controls not generally provided for making transmitter adjustments; 2)service difficult and expensive; 3)sometimes "tinny" loudspeaker sound, as result of more compact overall design.

Tube disadvantages: 1)circuit components age faster, due to inherent heat build-up within cabinet; 2)transceivers physically larger, table-model-radio style; 3)higher overall current drain – of particular concern to mobileers; 4)as base stations, usually most costly; and 5)lower resistance to shock and vibration.

At one time, any solid-state transceiver was regarded as exotic and "state-of-the-artish." Today, somewhat the reverse is true – particularly if you look at the construction of the most expensive, top-of-the-line base stations money can buy. To keep up with the Joneses these days, manufacturers are gilding the lily with all kinds of circuit extras for the solid-state jobs. Be wary.

The IC War

With the advent of the integrated circuit – particularly the linear IC – transceiver makers have been having a field day, racing to come up with the greatest number of chips per unit in a given price bracket. Entire advertising campaigns are being implemented totally around the space-age, wonder-drug, wash-day miracle.

Sound a bit skeptical? Okay. It is. In no other two-way radio communications service, marine, business and amateur included, are ICs getting as much attention as they do in CB. The reason? Except for CB users, communicators buy on the basis of performance. Designers, realizing that CB is as close as a radio service can come to a consumer market, are simply wiring in ICs for the sake of appeal.

Some sets have even been found to contain defective reject chips, wired in, but nonessential to the CB rig's operation. Recall in the early sixties when Japaense manufacturers were exposed by the FTC for calling a 21-transistor radio a 36, even though 15 transistors weren't even functional? Remember the number-of-transistors-per-radio war? Well, the same thing could happen in CB. Just make sure you're paying more for something of value and not getting a lot less performance.

Don't get trapped in this race, trying to out-top your buddy for nothing more than the sake of getting one more IC than he has. You'll pay through the nose for a dubious advantage.

Delta-Tune and Off-Frequency Transmission

Unbeknownst to most of us, our transmitter's oscillator circuitry has the built-in capability of pulling even the most accurate transmit crystal considerably off frequency – making it, of course, that much more difficult for a properly aligned receiver to copy the signal. The FCC says that the oscillator must be crystal-controlled. Yet, dependent upon circuitry, a channel 10 crystal in one rig may well be on channel 11 if plugged into another transceiver.

Furthermore, receiver and transmitter crystals are not interchangeable since they are cut for different frequencies.

A transmit crystal is often, for example, exactly one-half of the desired 27 MHz frequency; transceiver circuitry later doubles it. Receiver crystals, on the other hand, are often "third overtone" units that function quite differently and demand a totally different oscillator follower. And for equally obvious reasons, applied voltage is considerably higher in transmitter oscillators than it is in receiver oscillators.

Stray capacitances caused by nearby components, the crystal case itself, and a host of other variables can result in a "pulling" of your transmit frequency by as much as 2.5 kHz – about twice the frequency tolerance allowed by FCC rules and regulations.

Over the years manufacturers have made many circuit improvements aimed at lowering these differentials. Today it is very few sets that are off frequency enough to break the law. It is known, though, that of all CB transmitters on the air, fully 70% of them are enough off frequency to be noticeable.

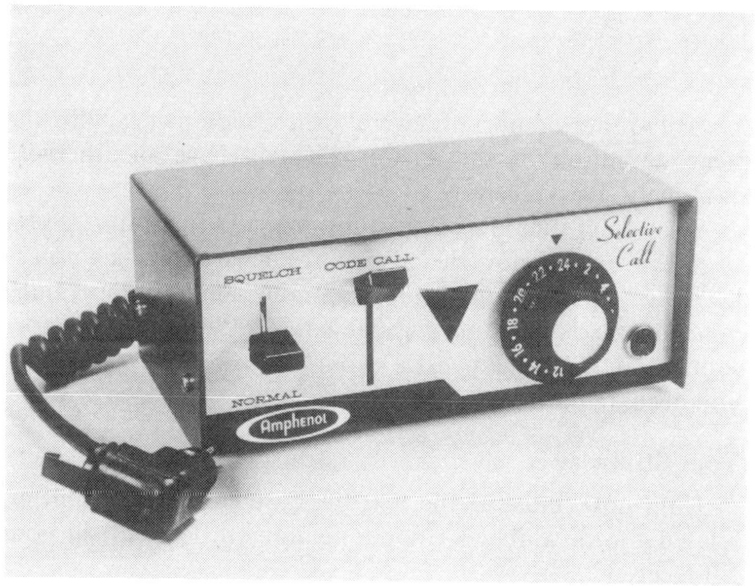

Here's a selective call add-on device. What it does is interrupt the squelched receiver only when it detects a decoded signal of predetermined tone sequence. To encode, CB'er depresses "Code Call" lever. Rotary selector at right sets any one of 24 individual code combinations.

For mobile use, compact selective call devices such as those above are available. To be able to use tone signaling, however, usually requires that you anticipate such use in purchasing a CB transceiver with accessory plug-in socket. Most manufacturers who provide this capability also sell compatible signalers.

Delta-tune, or one of several other trade names with the same definition, is the answer – if you have an otherwise extremely sharp receiver.

What delta-tune does is to allow you to "fine-tune" 3 kHz or so above and below the channel frequency to which you're listening. Operationally, it's much the same as fine tuning your television set. As a general rule of thumb, the more sophisticated (and expensive) your CB transceiver, the more you need delta-tune.

Your CB Receiver

Until now, most of the text has covered only transmitter-related topics. But what about features to look for in your CB receiver?

The old adage, "You can't work 'em if you can't hear 'em" is appropriate here. While CB transmitters are generally comparable – as a result of tight FCC regulations concerning frequency control and power – receivers are in no way in the same boat. There are underdesigned and overdesigned receiver circuits abounding in the marketplace.

At one time, the ultimate in a CB receiver was a tunable double-conversion configuration. Today, triple-conversion is obtainable for the right price. These two, incidentally, are contrasted with the conventional superheterodyne (single-conversion) circuit roughly comparable to that in your automobile AM radio. Adequate (the single-conversion) it is; highly selective it is not.

The only thing wrong with this line of reasoning is that not everyone requires a superselective receiver. As indicated in the section above dealing with delta-tune, there is such a thing as having a too selective receiver, especially if she lacks delta-tune and is strictly crystal-controlled, or "fixed-tuned."

Additionally, triple-conversion does not address itself to sensitivity – which can be extremely critical in outlying areas not plagued with extensive channel congestion.

Operating range is largely dependent upon the sensitivity of the receiver. Your station's operating effectiveness overall is directly dependent on the receiver's selectivity and ability to reject or limit unwanted noise. And there is a lot to be said, too, for the operating skill of the individual CB'er. Be that as it may, though, the receiver section of any CB set should bear extremely close examination.

Noise Considerations

Especially to mobile operators, the ANL, or automatic noise limiter, is one of the most critical "musts" in a CB rig. A number of circuits are used to achieve noise limiting, and of course some are more effective than others. This is another one of those items that has to be tested before you buy under actual operating conditions. If you own a Volkswagen or Ford, rest assured you will have more trouble than most with ignition noise. Compare several CB rigs, if possible, hooked in series to a common antenna to see which has the most effective noise limiting. Needless to say, your ANL's efficiency on a comparative basis should be judged by how intelligibly a weak station can be copied under severe noise conditions. An effective ANL should be one that makes a weak station's audio audible when otherwise (ANL switched off) it would not be discernible from the noise. You'll be surprised how widely ANL efficiency varies from one receiver to another.

Bear in mind that you don't want signal-to-noise ratio adversely affected much by the switching in of your ANL. Some insertion loss is inevitable, but the most misleading circuit is the one to avoid – the ANL that silences noise, but silences signals at the same time!

Noise Silencers. While most all CB transceivers are equipped with ANL circuitry, you'll expect to pay a bit extra for your noise silencer. What is a noise silencer? Quite simply, it's a means to prevent the noise impulse from reaching much beyond the i-f or rf amplification stage – if at all. What it does is create a "hole" in the signal corresponding precisely to the noise pulse being received. What you hear – for a split-second, at least – is nothing. Then the normal audio is resumed. These kinds of circuits are particularly effective in combating sharp noise spikes such as created by ignition noise and the like.

Fixed Tuned or Variable Receive?

While the trend today is definitely toward fixed-tuned reception with delta-tune trimming, there are still quite a few transceivers on the market offering continuous tune receive. What's the difference?

Operationally, it's somewhat comparable to the VHF vs UHF television tuner situation in which the VHF is essentially detent, the UHF continuous. The detent is far more accurate and fast, although most of us find ourselves "trimming" with the fine tuner anyway to correct circuit aging or misalignment that results in off-frequency performance. With UHF, however, once the station has been located, you've got a good picture – all in one operation. The only thing you have sacrificed is accurate calibration.

The fixed-tuned vs variable is a personal preference situation for the most part. While many users wouldn't dream of owning a detent-tuned rotary switch rig, their proponents counter that in a congested atmosphere such as 27 MHz operation in a metropolitan area you'll never know what channel you're listening to with variable tune. They claim calibration can't be depended upon and further point out what we've already discussed – that most rig's transmitters are somewhat off-frequency anyway.

The Spot Switch

Another item that is somewhat of a carryover from amateur radio, the "spot" switch permits the transceiver's tuning knob to be adjusted precisely for the transmit channel frequency. What happens is that the transmitter oscillator is switched on, permitting the carrier to be heard on the receiver without blowing out the front end with the full effect of a hefty five-watt signal.

This is particularly useful for those with minimum-coverage CB transceivers. The typical user will ofttimes forget which crystals he has in his transmitter. The "spot" switch will tell you fast.

Most spot switches consist of a normally open spring-return pushbutton mounted on the front panel of the transceiver. For anyone with a continuous-tone receiver, a spot switch is a must.

The Squelch

Most rigs offer some kind of squelch, a useful circuit feature that acts to eliminate background noise during

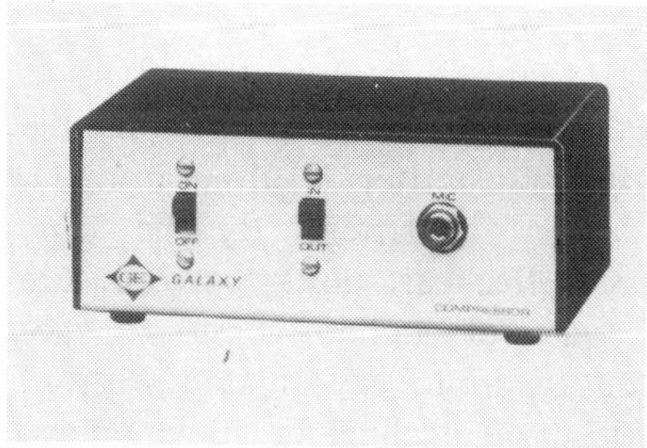

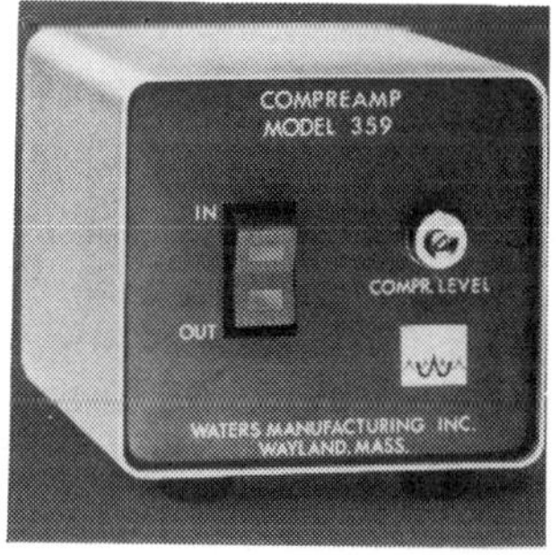

These two audio-speech compression devices can be extremely effective in providing additional audio "punch" to a CB transceiver. Speech clipping is particularly applicable when communicating over long, weak-signal conditions where signals are below S7 most of the time.

standby operation. A squelch is really an electronic switch that effectively cuts off all audio output from the receiver section, except when a signal that exceeds a preadjusted level acts to "open" the transceiver, allowing the audio output to be heard.

How Much Audio?

In most CB transceivers, the audio amplification system serves a dual purpose; 1)to act as a microphone speech amplifier on transmit and 2)as a receiver audio amplifier when not transmitting. Many sets include a third option for the same circuit: a public address system amplifier.

In nearly all units, the transmitter microphone amplification function is preset, while the receiver volume is variable. Frequency response is kept within the general speech range, or about 300 Hz to 2.5 kHz. If the response bandwidth were much higher it might produce an illegally wide transmit signal capable of bringing the FCC to your door in a hurry.

In a plate modulated transmitter, as most CB transceivers are, what is needed for 100% modulation is roughly one-half of the rated power input (rf) to the final stage of the transmitter. This translates to about 2.5 watts of required audio. For this reason, you'll find that most transceivers contain an audio amplification capability of 2–3 watts. This is also more than adequate to drive the set's loudspeaker.

Beware the transceiver that offers more than 3 watts of audio. If microphone gain is variable, it is quite possible to severely overmodulate the transmitter, resulting in much distortion and a good deal of "splatter." By the same token, watch out for Class D five-watt units that only are capable of generating 1 watt of audio. These sets are annoyingly undermodulated, cannot be corrected without major surgery, and will bring you nothing but unhappiness in lost contacts. Nothing is more frustrating than to try to pull intelligence from an S6 signal that suffers from 65% modulation.

Public Address and External Speakers

With the above facts in mind, appreciate the value of a public address system that boasts only 3 watts of amplification. But this isn't to say that a well-matched speaker attached to the public address terminals can't be useful. In small gatherings of people in a confined area, such audio is more than adequate. But for genuine public address applications – such as in auditoriums, rock festivals and the like – you'll need far more audio power.

Let us not overlook, however, the usefulness of an external speaker jack or terminals! For the mobileer on a camping expedition, for example, a typical solid-state transceiver can be left "on" for monitoring purposes for literally days without appreciably affecting the automobile's battery performance. With a remote speaker atop the car roof or facing the campers alongside the car, it can make the difference between hearing the call you've been waiting for, or not. Applications are as many as CB'ers have ideas.

An RCA hybrid linear power amplifier, this typifies what most CB'ers expect but don't get in most transceivers on the market: a real fistful of rf power. Instead many ICs currently offered only add to the equipment's cost while actually not improving performance over an equivalent solid-state transistor circuit.

This CB transceiver contains a public address feature the operator at left is using now. Unfortunately, this gear – like many – just doesn't have the degree of audio power needed to cover a large area in an out-of-doors situation.

If you anticipate a possible need for external loudspeaker, you'll find this provision isn't hard to find and at little or no price premium.

Receiver RF Amplifiers

Now that we have dispensed with audio amplifiers, let's take a look into the receiver circuitry, following the signal you're listening to. The lack of an rf amplifier may make your hearing the signal in the first place impossible. Similarly, lack of rf amplification increases your chances for image frequency detection and adjacent channel interference. In short, it could substantially decrease the overall performance of your receiver – affecting sensitivity, selectivity, the works.

To those of you without adequate rf amplifiers, there are a host of add-on preselectors and preamplifiers on the market that can be wired into your receiver by an experienced technician, with good results. Fortunately, most transceivers today have fairly good rf stages in their receivers – a few even offer an outboard rf gain control. This control is useful in situations where you want to detect weak stations by means of heterodyne (beat frequency oscillator), SSB or DSB stations, and also to prevent front-end overload from a powerful, nearby CB'er whose antenna is aimed right at yours.

Two types of rf amplifiers are found in today's CB equipment: 1)an untuned configuration and 2)a tuned rf stage. Untuned refers to an rf amplifier in which only one circuit in the receiver input is tuned. In reality, the circuit is more of an impedance-matching operation than an rf amplifier; most CB amps, to be truly effective, have at least two tuned stages – input and output. For best performance, the tuned input circuit should be entirely separate from the network of the transmitter. Such dual use of a transmitter's output circuit comes at the expense of receiver sensitivity.

In spite of considerable competitiveness among CB manufacturers to beat each other out in the specifications one-up-manship game, sensitivity figures, when determined under laboratory test bench conditions, are remarkably similar to one another. This is true at times even when rated sensitivity varies considerably.

As a guide, one microvolt (expressed as 1 μV) for a 10 dB *signal-plus-noise to noise* ratio is generally more than sufficient for most applications. Since mobile ignition noise can often be of intensities several times this figure, it obviously doesn't make sense to spend extra dollars for 0.50 or 0.25 μV sensitivity in most in-car installations. Other users, say, off in the country at quiet base stations, however, might find the extra sensitivity well worth the money.

Crystal and Mechanical Filters

Both mechanical and crystal filters are now available in sophisticated CB receiving equipment. While the mechanical types are generally considered to be somewhat more rugged, both mechanical and crystal units – which do add to the cost of the unit – do much in the area of increasing receiver selectivity.

What they do is sharpen i-f (intermediate frequency) bandpass to a predetermined razor's edge, allowing only those received signals falling precisely within the narrow passband to be heard. Since congestion and off-frequency operation are becoming the vogue, this can be useful in separating the desired signal from the host of causes of interfering noise – even though all the signals being dealt with are allegedly on the same CB channel.

As soon as your filter is switched into the circuit (in most cases, an outboard control is provided), you'll notice a distinct change in audio quality . . . generally for the worse. In most instances you'll find it desirable to leave the supersharp filters out of the circuit altogether. When congestion piles on, however, the filters can make the difference betweeen copying the desired signal – albeit at some deference to audio quality – and simply losing him in the screeches on channel.

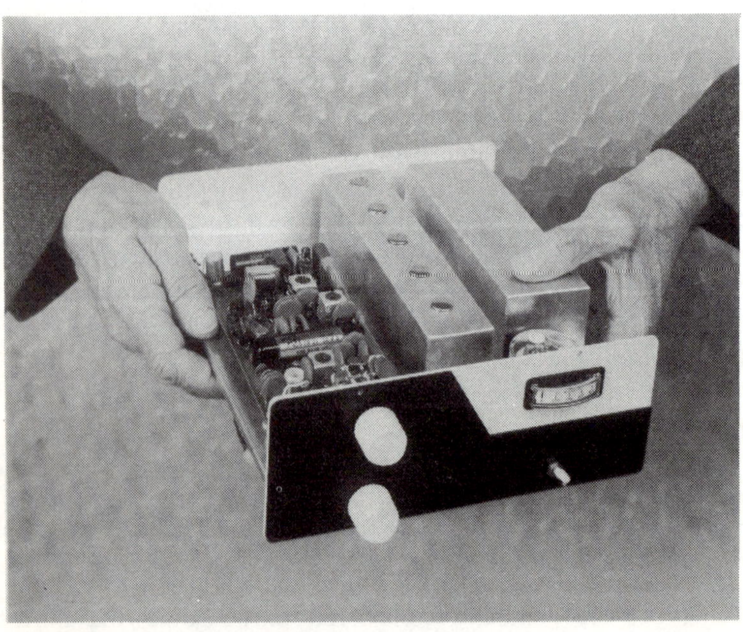

Large tubular component running horizontally on the circuit board (above) in this transceiver is a Collins mechanical filter, topping off a double-conversion receiver for increased selectivity. In many CB rigs, the filters can be switched in and out of the circuit, as needed.

Wrap-Up

There are a lot of features, most obvious, that we have elected to omit. Such things, for example, as physical size and panel layout, provision for microphone mounting bracket, accessory cables, and the like are strictly personal items. Other circuit performance features you'll want to look into yourself.

That a prospective equipment purchaser should operate the gear before buying cannot be overstressed. From a design standpoint, the transceiver might be the world's most advanced rig in the country – but if you find the receiver audio tinny or some other basic function annoying to you personally, you'll regret having spent the money. The big mail-order houses probably have a lot of customers in your community. Seek them out, ask to see the equipment, then – and only then – send off your money.

But most important: Determine your *own* CB equipment needs carefully. Make a list of features you feel are "musts," splitting them into tertiary levels. Satisfy the most fundamental of your requirements before matching any of the "extras" to your second or third group on your list. You'll be glad to took the time to examine your requirements – instead of buying strictly on emotion. And you'll be leagues ahead of the pack in really getting your CB dollar's worth.

Getting a CB License

Leo G. Sands

Getting a CB license should be as easy as getting a driver's license in a state where nó driver examination is required. There are no tests to take and only one license is required for covering any number of transmitters. But some people make it difficult because they don't "read" what it says on the license application form. Some people don't read it correctly (and understandably, because the form was prepared by lawyers), but it's easier to read than an income tax form.

If you fill it out correctly, you should get your license in four to six weeks. If you make an error, your application will be returned to you and you'll have to amend it and wait even longer.

You must have a valid Class D station license in the Citizens Radio Service before you can go on the air with a CB transmitter. You may not use someone else's call sign. But if an immediate member of your family residing with you has such a station license, you may – with the licensee's permission – operate one of his or her transmitters. Or, if you buy your own mobile unit, the licensee can assign you a unit number and you can then operate the CB transmitter under his or her license. However all such operation is under the control of the licensee who is responsible to the FCC.

If you go on the air without the proper license you can be fined heavily or even sent to jail, if you are caught by the FCC. However, you may operate a transmitter in the Citizens Band without a license if you use low power equipment (less than 100 milliwatts input power) which meets the technical standards of Part 15, FCC Rules and

Regulations, but licensed CB stations may not lawfully talk to you, since they are permitted to communicate only with "units" of their own stations and other "licensed" CB stations. Later, when you get your license, you can use the low power equipment for communicating with licensed CB stations, provided it meets the more rigid technical standards of the CB rules (Part 95).

You can get a CB license even if you don't own any CB equipment. If you're contemplating becoming a CBer (citizens band operator), it's a good idea to apply for your license before you buy equipment so you won't have to wait so long for a license after you get equipment, but you don't have to possess or apply for a license before buying equipment.

Are you eligible for a CB license? You must be a citizen of the United States, at least 18 years old, and must not have possessed a CB license which has been revoked by the FCC. If you are an alien, take up short wave listening. If you're under 18, you can get a ham license of your own. Or your father or mother can get a CB license and authorize you to operate a CB transmitter. In the latter case warn him or her that he or she will be responsible for how you use the transmitter and might have to pay a fine if you do wrong.

To apply for a CB license, you need a copy of FCC Form 505, which is available at FCC field offices and at most CB equipment dealers, and by mail from the Federal Communications Commission, Washington DC 20554. The form consists of a worksheet, the formal application form, and instructions. Also, there is a coupon for ordering Volume VI, FCC Rules and Regulations, for $1.25 from the Government Printing Office. Order it and read it. Volume VI contains Part 95 which governs CB. Every CB licensee is required to possess and read these rules.

Fill in the worksheet with a pencil. Later, copy the information on the formal application with a pen or typewriter. Here's what to put on the form at the numbered items:

Item 1: Your last name on the top line and your first

name and middle initial on the second line.

Item 2: Leave it blank.

Item 3: Your permanent mailing address in full, including ZIP code.

Item 4: Check the "individual" box.

Item 5: Check the "Class D" box.

Item 6: Check the "NO" box if this is a new application.

Item 7: Check the "NO" box if you do not now have a Class D CB license.

Item 8: Watch this one. On the blank line you are supposed to list the number of transmitters "to be authorized" by the license. It does NOT mean how many you possess or plan to buy initially. If you write in "one," expect your application to bounce unless you enclose a letter explaining how you propose to use "one" transmitter.

Why? The FCC will suspect you want to be a CB hobbyist who buys "one transceiver" for the sole purpose of "making like a ham" on CB, which is unlawful. There's nothing wrong with owning only one CB transceiver if you use it correctly.

If you write "two" or any higher number on the blank line, you won't be suspect. But if you plan to use only one transceiver and want to be forthright, enclose a letter with your application explaining why, as for example . . . "I plan to install a citizens band transceiver in my car so I will have means for calling for road assistance when needed" . . . or . . . "I live in a rural area where telephone service is unreliable (or not available) and I need means for communicating with others in cases of emergency."

Most CBers own two or more transceivers. What number you indicate at item 8 is the number of transmitters you can lawfully operate. If you make the number too small you will have to apply for modification of your license when you acquire more transmitters than authorized.

Note that there is no reference to base stations, fixed

stations or mobile units. Under a Class D license, all transmitters are classed as "mobile units" whether used at fixed locations or in motor vehicles, boats or aircraft, or if they are walkie-talkies or packsets.

Item 9: This one must have been written by a Philadelphia lawyer. How are you supposed to know if your transmitters are crystal controlled and/or if they have been "type accepted" by the FCC (or what all those words mean)? By all means put an "X" in the "YES" box. You can assume "Yes" is an honest answer if you use factory-made CB equipment sold by a reputable dealer. If you say "No," your application will be rejected.

Item 10; Check the "YES" box unless you plan to lease or borrow your equipment. In that case list the name of the equipment owner at 10(B) and check the "YES" box at 10(C).

Item 11: Check the "YES" box. You're supposed to have read and understand the CB rules (Part 95). If you indicate "NO" your application will bounce and you should not operate a CB transceiver.

Item 12: Check the "YES" box. By doing so, after signing the application, you will have certified to a federal agency that you will not violate specific CB rules (hobbying, skip, etc.).

Item 13: Check the "NO" box unless other than you, members of your immediate family living with you, or your employees are to be allowed to operate your transmitters. If you indicate "YES" explain "who else" in your letter of transmittal.

Item 14: If you are a U.S. citizen check the "NO" box. If you aren't a citizen, forget about a license.

Item 15: If you are a spy or a lobbyist for a foreign government, check the "YES" box and don't file the application. Otherwise, check the "NO" box.

Item 16: If you have been found guilty of a crime and were given a jail sentence of six months or more, or fined $500 or more (even if suspended), mark the "YES" box and forget about being a CBer or even a ham. Otherwise, check the "NO" box.

Item 17: Check the "NO" box if you are over 18 years old. If you aren't 18, don't bother applying for a license.

Item 18: If you indicated a P.O. Box or RFD number at Item 3, explain where the FCC can ordinarily physically find you.

Then turn the form over and read the statements under "I CERTIFY THAT . . . " and sign your name on the signature line, note the date and put an X in the "individual applicant" box.

Check the worksheet carefully. Ordinarily you should have indicated the following at the YES – NO boxes:

Item 9 through 12: YES.
Item 13 through 17: NO.

Now copy the information on the formal application form in ink (or typed), sign and date it and mail the application to the Federal Communications Commission, Gettysburg PA 17325, enclosing a check or money order for $8.

Items 2, 19 and 20 should be left blank if you are applying for a license as an individual in your own name.

If you want to apply for a license in the name of a business or a club, fill in Item 2 if the business is a proprietorship or partnership, and on the top line of Item 1 note the name of the business or club. If the business or club is a corporation, also fill in Item 19, and in the case of an unincorporated association or club, fill in Item 20.

If any partner or any director or officer is an alien, don't apply for a license. Although alien hams from certain countries can operate a ham station in the United States, because of reciprocal agreements, all licensed CBers must be U.S. citizens. A U.S. CBer can get authority to operate CB transmitters in Canada, but a Canadian may not do so in the U.S.

At the time you apply for a CB license, also order a copy of the rules, using the coupon furnished with the application form. Before you get your license, read the rules so you will know how to operate a CB transmitter legally.

When you get your license, it will authorize you to do the following :

(1) Communicate between your own mobile units and base stations on any of the 23 CB channels.

(2) Communicate with any other CB station on only Channels 9, 10, 11, 12, 13, 14 and 23, but not over distances in excess of 150 miles.

The license does NOT authorize you to communicate with unlicensed Part 15 stations, nor does it authorize you to communicate with ham stations, or other non-CB stations except in emergencies involving immediate danger to life and property.

After getting your license, which is good for five years unless revoked earlier, don't lose it by being caught violating FCC rules. For detailed information about how to use CB equipment properly, buy copy of "101 Questions & Answers About CB Operations" at your CB dealer or through a mail order house.

The CB Receiver–What You Need To Know

Robert M. Brown

With Class D CB transceivers beginning to look more and more like carbon copies of each other, how does a knowledgeable operator differentiate between the good and the bad? One place to start is to examine receiver specifications, since receiver circuitry – unlike the transmitter – isn't limited in design by FCC regulations.

Receiver specifications aren't nearly as complicated as they might first appear. The important thing to look for can usually be grouped into one of the following three performance "musts:" 1) selectivity, 2) sensitivity, and 3) effects of noise. Once you have an understanding of these three items, nearly everything you read can be grouped into place. Let's take a look at these key performance criteria.

Selectivity

What's so significant about selectivity? Well, this feature is generally conceded to be a receiver's ability to reject signals on frequencies other than those to which your CB rig is tuned. Class D 27 MHz CB channels, for example, are spaced as close as 10 kHz apart. What this means, then, is that your CB receiver – when tuned to channel 11 – should not accept signals from stations transmitting on the two adjacent channels, 10 and 12. This includes the upper and lower sidebands, if the signal happens to be single sideband.

Does CB receiver selectivity have an effect on range? You bet! Let's hypothetically suggest your receiver is tuned to CB channel 13, or 27.115 MHz. If no signals are detected on the channel, your receiver's automatic volume control (AVC) circuit automatically sensitizes the receiver so that the very weakest, distant stations will be audible. When an S5 signal appears on channel 13, the AVC desensitizes the receiver volume to keep you from being blown out of your chair. But here's the important point: Should a very strong, S9+ signal suddenly appear on either channel 12 or 14, it is entirely possible your AVC may desensitize. It could well make the difference between your hearing that S5 signal on 13, or not. Obviously, then, a selective receiver is desired.

There's no question that extremely selective receivers are available. They can be designed to reduce adjacent channel interference and desensitization of the receiver by strong adjacent channel signals. Usually, you can find these by looking through the specifications for "double-conversion superheterodyne" receiver configurations. In fact, triple-conversion receivers are available, too, for the most exacting receiving requirements.

A word to the wise, however. As selectivity narrows – improving apparent performance – transceiver cost sky-

Distributors like Electronic Wholesalers (this showroom is in Florida) frequently offer real bargains in reconditioned and used CB transceivers that only require a few minutes alignment to return them to original sensitivity performance.

rockets. In most instances the old adage that you get what you pay for holds true.

Why are the most selective receivers usually base station units? One reason is that as selectivity design is optimized, local oscillator drift becomes extremely critical.

If you're not in a hurry, even radio-TV repair shops like this one in Chicago will be adequate to do an alignment job on your pooped-out CB receiver. Prices are usually quite reasonable, if you can supply your owner's manual with circuit schematic.

And this is something not limited entirely to tube-type base stations. Another reason is that signals are liable to be stronger on adjacent channels more often at the base – you simply hear more overall – than in the mobile.

In the end, it becomes apparent that some kind of compromise must be reached. If in fact selectivity affects range, then selectivity is desirable in the mobile as well. If, however, stability becomes more critical, sheer vibration to the set in the mobile may wreak havoc on a super-selective receiver. Lastly, evaluate cost. Do you want to spend $400 for a super-selective receiver in your specific application?

Sensitivity

We said there were three equally significant parameters when evaluating CB receivers. While selectivity is important, the sensitivity of your receiver is even more directly related to your overall communications range. The saying that "You can't work 'em if you can't hear 'em" holds a lot of water.

Stan Streeter, a serviceman for Store Fixtures Sales Company in Buffalo, New York, frequently has his radio equipment checked out by a professional technician to insure best possible reception. Here, he's checking in to determine whether any new jobs have come up.

Since the power of Part 95 Class D citizens band transmitters is generally about 3.5 watts to the antenna, most CB signals are considerably weaker than those radiated by broadcast stations. As you know, most metropolitan television/FM/AM broadcasters travel with power inputs to the final stage of their transmitters of in the order of 50,000 watts. The FCC limits CB power inputs to five watts. Obviously, an extremely sensitive receiver is required for optimum results.

Professional two-way radio communications service centers like Alec's Radio in Massachusetts are happy to check out CB receivers for a modest fee.

How do you measure CB receiver sensitivity? The most commonly accepted means – shown in most spec tables – is in terms of microvolts. These are abbreviated as "μV." A CB receiver with a rated sensitivity of five microvolts, for example, will produce a predetermined increase in audio output when modulation is applied to an otherwise unmodulated five microvolt CB signal applied to the antenna input connector of the transceiver. To be able to respond adequately to a CB signal of only five microvolts – which translates to exactly five millionths of one volt – the CB receiver, obviously, must be well designed.

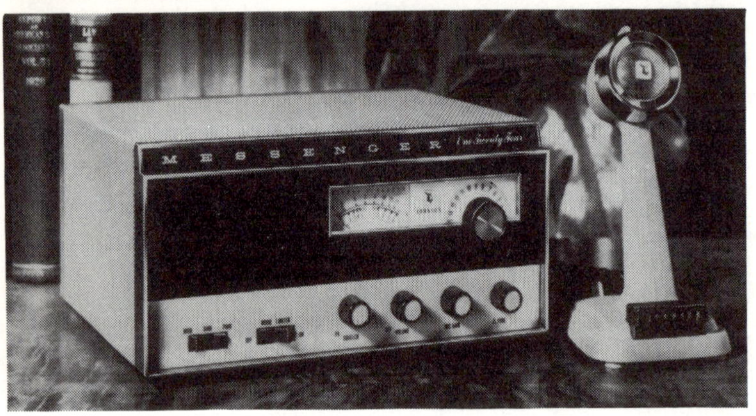

Most of what you pay extra for in the more expensive CB base station transceivers is superior receiver circuitry that offers 1) optimum selectivity, 2) good sensitivity, and 3) minimal noise level. E. F. Johnson Messenger 124 above features tunable receive, adjustable squelch, front panel noise limiter.

Sensitivity is one of the first things to drop off as a transceiver ages. Components tend to heat in time, affecting performance. Alignment tends to "creep" from where it should be – often as a result of sheer physical vibration or jarring. Periodically it makes sense to have your receiver realigned by an expert technician. You may find your five microvolt sensitivity has deteriorated to 20 μV!

Often an alternative to a CB'ers search for the ultimate in sensitivity – perhaps 0.15 μV – is the addition of an inboard or outboard preselector or rf preamplifier.

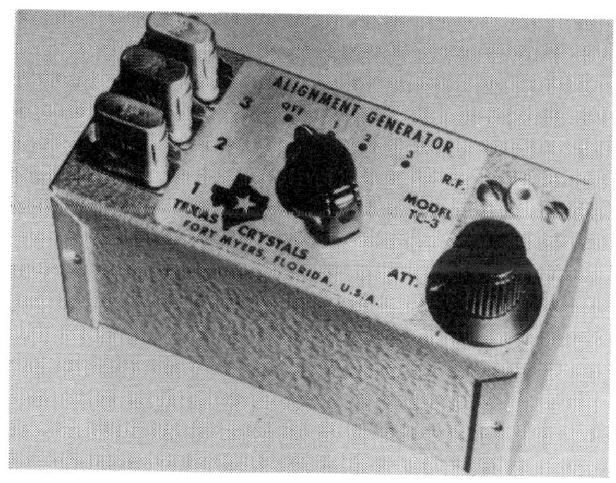

Alignment generators like this one are generally used to optimize receiver sensitivity by service technicians.

These devices, available at reasonable costs, can be wired into the CB transceiver betweeen the receiver antenna input and the first receiver stage. Called "boosters" in most cases, they are simple, relatively foolproof, and worth every dime of their cost. They are, of course, most effective on single-conversion under-$100 CB transceivers. They are particularly effective on the bargain transceivers (under $60) commonly found on resale counters at CB distributors.

Noise

We said there were three primary elements to be considered. Last but certainly not least is noise. While your CB receiver may have a rated sensitivity of 1.0 microvolt, its useful sensitivity may not be very high at all because of noise transients.

As far as your receiver design configuration is concerned, noise – of just about any kind – is a signal. If your noise level is high, the AVC circuitry will automatically attenuate the receiver's gain in exactly the same fashion as it would if it were receiving, say, an S9 CB signal.

As a result of all this, noise that produces a 16 μV "signal" at the CB receiver's antenna input connector will

Gilvin Roth YMCA in North Carolina specializes in summer camp training for CB'ers wishing to learn more about how their sets operate and what to do when they don't. In "A" above, CB students learn how to read their receiver's circuit diagram. In "B" same students study to prepare for amateur radio license examinations.

attenuate the receiver gain and make it virtually impossible to hear a 3 μV CB radio signal. When the noise level produces, say, an 8 μV signal at the antenna connector, a CB radio signal – to be heard – must be of greater intensity, or, say 9 μV. In all cases the CB signal must be strong enough to override the noise.

So what's with all the μVs? What about dB? Well, the good two-way conversation on CB is probably an exchange of signals that ride somewhere around 48 dB. This

Here's a miniature rf amplifier, designed specifically for in-board mounting in CB trans-ceivers. Powered entirely by the CB receiver's internal volt-age, the preamplifer can add several "S" units of additional signal strength above noise level. These are especially ef-fective when installed in older, or less expensive base and mo-bile CB rigs.

translates into something that looks to be a voice level about a milion times greater than the level of background noise. In most CB receivers, this signal-to-noise ratio at the maximum rated sensitivity is somewhere around 6 dB.

Your ' S" units are approximately 6 dB apiece. In other words, an S7 signal would be seven times six, or 42 dB. If the noise level were 6 dB or one "S" unit, then the signal would be said to be 36 dB or "36 dB above the noise." If your S-meter were set so that the normal noise level was "0," the reading would be S6.

For the curious, here's some more information on the subject. Let's say that an ac voltmeter is used to measure the actual signal at the receiver's antenna input terminals. Suppose the background noise measured 0.2 volt when a 2 μV unmodulated signal was received. Now, if the CB receiver's output voltage should double – to 0.4 volt – when the AM signal was modulated 30%, the receiver could then be said to have a measured sensitivity of 2 μV for a 6 dB signal-to-noise ratio.

What happens is that the demodulated output of the 2 μV signal makes the CB receiver's audio output voltage increase to double the level of the background noise. The stronger audio output signal, of course, contains both the noise and the modulated "talk" – but for reliable communications, you'd probably want that signal-to-noise ratio to be as much over 6 dB as possible.

Remember All Three

When reading receiver specifications, or checking out reconditioned and new CB gear before purchase, force yourself to evaluate 1) selectivity, 2) sensitivity, and 3) noise rejection of the set in question. They can't be looked

Noise level can vary from one location to another. Here, range is inhibited not only by tall buildings, but also ac line transients, overload from nearby broadcast stations and high degree of vehicular ignition noise.

This installation – CB on a farm tractor – will test the noise-limiting circuit of even the best transceiver. For such an application noise rejection (from the tractor engine and ignition) will be far more important than the receiver's ability to effectively reject adjacent channel signals.

at separately, but rather should be weighed against one another.

If possible, check out the prospective CB set in your own shack or auto. If, compared with other transceivers, it shows an extremely low noise level while receiving CB signals at the same general signal strength as the other sets, you might want to overlook a sensitivity rating that isn't quite up to par on paper. Noise rejection can be the difference between copying a signal and not.

If you live in an area – and I don't imagine there are too many like this any more – where you won't be plagued by pileups and crowded S9+ signals, perhaps you can afford the savings realized in purchase of a less selective CB rig than your city brethren might be able to get away with.

In any case, buy in terms of your specific needs and you'll get far more from your receiver than you dreamed possible!

Simple Short Wave Receiving Antennas

Paul Schuett WA6CPP

An antenna is essential to any radio receiver. Antennas may be the compact ferrite loop or may be an elaborate longwire suspended high above the ground. Although generally speaking, the longer and higher the antenna the better the reception, many quite simple antennas can be used with surprisingly good results. Let's look at some simple receiving antennas that can be easily installed by the beginning or experienced SWL/BCLer.

The simplest antenna is a ten to fifteen foot piece of wire. Connected to the antenna (or antenna 1 with antenna 2 shorted to ground) terminal, draped across the floor or along the ceiling, this wire will bring in considerable DX.

I used four versions of this antenna in my college days. Space was always at a premium, so the antenna had to be kept short. For some time the wire was run along the floor; after a while it was revised and run along the ceiling, through a transom, across the hall, through another transom, terminating in another fellow's room (nice of him). The next year, the wire was hung out the window from the second floor; finally, with the cooperation of the people upstairs, the wire was run up. The wire was always about fifteen feet long.

Generally speaking there was no appreciable difference of reception qualities on the four arrangements. BCL DXing was carried on, as was SWLing, with the more popular countries logged in consistently.

Probably the best of the four arrangements was the one that went upstairs. This had the advantage of height, but the disadvantage of being near the building with the fluorescent lights, electric shavers, and other assorted noise sources.

Several possibilities are wide open for the DXer at home. If you have an attic, regardless how small, an antenna wire can be stretched end to end. For best results, terminate the end on an insulator – a TV standoff works fine, an egg insulator, or a regular antenna insulator. I use Johnson 136-104, priced at 30¢, which has superlative qualities – high tensile strength, good insulation, and low price.

The attic antenna can be a dipole – center fed with balanced line, or end fed with coaxial line. The center-fed dipole works quite well on one frequency band, or odd multiples of that band and is reasonably directional, receiving signals best broadside to the wire. The lead-in line should be balanced wire, preferably 75Ω cable, available from the various supply houses. The other end of the lead-in attaches to antenna 1 and antenna 2 terminals on the radio. If your receiver does not have the two antenna terminals, it would be better to use the end-fed antenna, although the second wire could be grounded.

The end-fed antenna has an advantage of not being frequency-selective if it's long enough. A long wire is also less subject to fading. Although signals do come in best broadside the wire, the longwire demonstrates more nondirectional characteristics, especially when the wire gets to 75 feet or longer.

The coaxial lead-in should be RG-11/U or RG-59/U cable. The RG11 is much larger and more expensive. There is considerable discussion about the comparative advantages of the two. Both demonstrate identical electrical characteristics, but the RG11, being larger, would have less resistance. For long runs, the 11 is preferred; shorter runs, the 59 is fine. Consider your budget and talk to your friends. The advantages and disadvantages of large and small cable generate considerable discussion among electronics people.

The center wire is the "hot" wire and is connected to the antenna wire; the shield is grounded at the receiver.

Coax can be run through conduit, buried, tied to a mast, and used about anywhere. Balanced wire (such as 300Ω television twin-lead) has to be kept away from metal objects that might disturb the electrical currents.

If you have a 30 to 50 foot mast atop the house for the TV antenna, consider using the TV lead-in wire as your antenna. Connect both wires to antenna terminal 1 and you have a vertical antenna. I have used this method for several months with excellent results, logging Bulgaria, Red China, Tokyo, as well as the stronger regular signals, without even trying for the rare ones.

The standard TV mast can be used as a vertical receiving antenna. The existing mast can be used, sharing with TV reception, or another can be erected easily. There is a possibility of interaction between the TV and radio, especially for the discriminating listener.

For optimum results, use nylon rope as guys, or break up the metal guy wire into small sections with egg insulators – maybe into 6 or 7 foot sections. Have the mast set on an old inner tube or other insulating substance. Connect the lead-in right to the bottom of the mast, making sure you clamp securely. Any type of coax will work fine for this, although I would prefer RG8/U or RG58/U.

If you like to listen on VHF, a groundplane could be installed at the top of this mast. Since the radials of that antenna are grounded, isolation should be provided between the two. If you only listen to one at a time, this is not so important if the VHF coax is disconnected; and in some installations you may not be able to tell the difference at all.

Using another section of the same inner tube the mast sits on, wrap this around the part of the mast where the groundplane will clamp; then run the VHF coax down the inside of the mast.

Another variation would be to connect a wire at the bottom of the antenna and have it run straight down from the mast into the attic, between the walls, or whatever, increasing the effective length of the antenna.

If you have a lot of room, erect two 50-foot TV masts outside, using stakes to anchor the guy wire, and having the mast sit on a large 10 x 2 plank about 2 feet long to distribute the weight. Connect a pulley at the top of each mast and string nylon cord through the pulley, having the cord long and strong enough to reach the ground and hold a long wire in the air. Suspend an antenna between the two masts, using antenna wire such as Belden 8008 or 8009, with a Johnson 136-104 insulator at each end. Feed one end with RG11/U or RG59/U and you'll pull in anything that's there. I have such an arrangement with the antenna 50 feet high, 125 feet long, using about 200 feet of RG11/U coax buried between the mast and the radio shack. I pick up stations in Hawaii and Florida stronger than the local broadcasting station.

Lightning protection is a must in any part of the country. In California's central valley we have few electrical storms; consequently some folks neglect proper lightning protection. The one time you are hit will make one worth while. I have three runs of coax into my operating room, each run protected by a coaxial lightning arrestor. Although these units are a bit expensive, the protection of your equipment should be the first thought in your mind.

Any coax connectors outside or buried should be weatherproofed, as should the coax lightning arrestor. I bought some "icky goo" at the hardware store that I smear on the connections. When dry it turns into a plastic coating. Moisture inside the coax can cause problems and the time to prevent problems is before they start. I have one coax junction buried. This was carefully taped with 3M plastic tape, and then two or three coats of this plastic protection stuff were applied. All exposed coax connectors are given generous treatments of this compound.

A good antenna system will work wonders even with an inexpensive receiver. I would rather have a good antenna and a mediocre receiver than the world's best bandspreader with a mediocre antenna. One of these ideas is probably ideal for your location. Try one or two of them – and listen to the DX.

The Radio License

T. W. Laffin W1FJE

A radio license is permission to operate a radio or television station, to broadcast, or to communicate with another station. It is for a number of other purposes including but not limited to experiments with the science. In the United States no permission or license is needed to receive any communications, and there are no licenses or approvals required for equipment used only for receiving.

In this nation and in many other countries all over the world, the license is permission to transmit communications. Agreed at Geneva were radio identifications for each place on earth which would be transmitting radio communications. Every country has a different identification, and every station within each country also has a special identification. In the U.S. there are hundreds of thousands of licensed radio or television stations. These licenses are issued by the Federal Communications Commission in Washington. There are different types of licenses for different types of communications. One of these is the group of Amateur Radio Service licenses. Of the different classes of amateur licenses the most popular license is the General class.

The Amateur Radio Service General class license is permission to transmit radio or TV communications. There are regulations in the U.S. on operating the station, and the General class license allows the person to transmit far enough to be heard anywhere on earth or in space as clearly as the equipment the operator builds or buys will perform. The General class license authorizes many frequencies which the operator can select to use, including 160, 80, 75, 40, 20, 10, 6 and 2 meter bands; from

220–225 Megahertz; 420–450 MHz; 1215–22,000 MHz (in five bands), and all above 40,000 MHz.

The Technician class license authorizes all amateur privileges in the U.S. bands above 6 meters, or 50 MHz; the Novice class license authorizes the use of the 80, 40, and 15 meter bands. The Novice license is the first, or beginner's, license. It gives permission to operate for two years, and cannot be renewed. All other Amateur licenses can be renewed every five years. On the 80, 40, and 15 meter bands the Novice can operate Morse code, but with less power than the General class license allows. The application for all Amateur licenses includes an examination by the FCC.

There is one kind of radio license in the U.S. which does not require any examination of the operator's ability to use and understand communications. This kind of license is called the Citizens Band license. It is not a type of Amateur license, but its purpose includes very limited use of the kind of communications authorized by an Amateur license, and with a very low transmitter power which cannot exceed 5W input to the final stage. Any citizen of the U.S. (18 or over) can apply for a Citizens Band license and operate a two-way radio station, fixed, portable, or mobile. As with all licenses, even the low-power Citizens license requires a station identification. The FCC also approves the equipment used under the Citizens Band license, so that no inferior equipment will acccidentally or intentionally be used which would interfere with other communications. Since the Radio Amateur operator must pass an examination, no FCC equipment approval is required for Amateur Radio equipment. Part of the Amateur exam includes also the instruction to operate within the accord of the Geneva Convention. This is so that each operator who is transmitting outside the U.S., or to other radio operators in the world, will know what the international agreements are and abide by them. Since the Citizens Band license will not likely provide for operation that could be heard for more than a short range, and not likely outside the U.S., no examination including the

Geneva guidelines is required. However, upon application for a Citizens Band license, the operator must be able to prove he has a copy of the FCC laws which govern the Citizens Band Radio Service, and he must have read and understood these laws. Other than the Amateur license, each operator and each station must be licensed if transmission is intended. The Amateur license is a combination of both operator and station license.

Other FCC licenses include the First or Second Class Radiotelephone license, the Broadcast license, and the Restricted Radiotelephone license, to mention the most common. The operator or holder of a First or Second Class license has permission to control or change the operation of broadcast stations, which the Amateur licenses do not provide for. The broadcast stations must have an authorized licensed person at the transmitter all the time, to make sure the transmitter is operating properly. The examinations required for Second or First Class licenses are much involved in the technical knowledge needed to design, maintain, or operate a radio station. The Restricted Radiotelephone license is a permit, or permission to transmit on special bands, such as the bands used by airplanes or ships and small boats, for example.

Except for special times, such as the Apollo program, or for research, there is only limited use of the bands above 2 meters, or 144 MHz, which can be heard with special radio equipment around the world. Another use of the higher frequencies includes the communications satellites, which carry television from one continent to another. For example, the English can see an event in the U.S. on BBC television at the same time the event is happening, or vice versa.

A new use of the 420–450 MHz Amateur band is in Amateur Television (ATV) communications. Some U.S. and other amateur operators have built their own television stations, but their range, even with one kilowatt, has been limited by the characteristics of the higher frequencies. However, Amateurs (or "hams' , as we are called) have been experimenting with their ATV transmitter antenna pointed at the moon. This means that the transmission "bounces" off the moon, returning to someplace else on earth; it is called E.M.E., and it has been a successful experiment. Some hams have even built and are using two-way color television, in addition to radio communications. E.M.E. is an opportunity for nearly anyone with the proper license, anywhere in the world, to communicate with another person – some day even with color TV.

The reason for licenses is not as much for control, but because there is a need to provide for the least interference between stations or services. When two or more stations are operating on the same frequency, near each other, or with a certain power, they automatically "jam" each other. With the wave spectrum divided, the stations identified, and frequencies or bands allocated for a variety of purposes, the purpose has been to avoid "jamming." Other than obeyance of the laws of decency and respect, there are no regulations for what is said or shown in the transmissions originating within a free nation. Licenses can be revoked if they are misused in the matter of decency and respect for all.

Even with a Novice class license, an operator can communicate – and prides himself in communication with the fifty states, as well as many European and other countries. The Citizens Band license, and the good experience gained with the use of the permission to operate, provides an easily-obtained familiarity with two-way communication which can lead an original interest in radio into higher-powered equipment under other licenses, or into other services or occupations.

Communications can effectually "pull" people together, as well as set people apart at safer distances. It began as a hobby, a strange and interesting science, and its path has followed primarily a course of public and private service, with a colorful and heartwarming history in the over half a century since its national and later international recognition as a very important mutual science. Every overseas or inter-country amateur radio operator from every country in the world is an ambassador for his nation and his hobby. Every SWL station is his audience, and there are many more SWL stations every year everywhere on earth. Some licensed operators spend more time listening than transmitting, and to you your speaker is an arena, as is the SWL – but with opportunity to participate actively at the time, and to comment to others; to enjoy and enjoin with mutual interest.

Of all international hobbies and professions, communications is among the strongest unions of brotherhood; the goals and ambitions of different nations, with a variety of reasons for transmissions, and by a variety of people, are all similar within communications and are protected by the phenomenon itself.

Pioneer Radio on the Prairie

Dr. W. C. Hess W6CXC

The scene is eastern Montana, circa 1926. Seated before a rolltop desk on which is stacked a small but complete broadcasting station and engaged in winding a hand-cranked phonograph is E. E. Krebsbach, entrepreneur of Radio Station KGCX, then located in Vida, Montana. The phonograph was vital to the operation of the station, its broadcast music being produced by the simple expedient of placing its microphone (which formerly saw service as a desk-stand telephone) in front of the Victrola.

Vida, a hamlet with a population then and now of only about fifty people was, in 1926, not blessed with electric power lines so the 7½W signal of KGCX was generated with the aid of a 32V farm light plant.

Farmers of the area, their day's work done, would fire up their crystal sets or perhaps a squawky Grebe regenerative receiver to listen to the evening's program on "that banker feller's station over in Vida." Reception was mostly via headphones except for those rare occasions when (wonder of wonders) relatively distant CKCK located in Regina, Saskatchewan, achieved loudspeaker levels with their signals. Such an event caused businessmen in the towns of the area to hurry home to bring their wives and children "downtown" to hear this marvel of the Twentieth Century. The trip to Main Street was made at breakneck speeds in Model T Fords with much dust generated from the unpaved streets, lest the signal fade out before the entire family heard it (invariably their comment was "It was coming in RIGHT ON THE LOUD SPEAKER") and because in those days, there usually was only one receiver in each village, more often than not being owned by the town physician or druggist, as befitting persons representing Science in the community.

The operator of the radio receiver sat amidst a farrago of wires, dry batteries, wet batteries, gooseneck speaker, huge loop antenna, etc., and was regarded with considerable respect in the community for his ability to snatch music and voices out of the air. The reader should understand that just anyone could not, of course, and would not be permitted to operate the radio receiver. One had to be qualified as a RADIO EXPERT. Many dials and knobs had to be manipulated precisely to bring something intelligible out of the squawks, squeals, and static crashes produced by the receiver, and if the expert was out of town, the radio was mute until he returned. The townspeople stood around it when it was operating, listening on multiple pairs of headphones all plugged into the same jack.

Telephones of the hand-cranked variety were commonplace in the villages of the area by the mid-Twenties, and the Montana mind could comprehend how talking across miles of wire was accomplished. The process of radio broadcasting, however, was not so easily understood. It was something tinged with wonder, with sheer magic, with unbelievability that a man could speak into a microphone and *be heard miles away without benefit of any connecting wires between the two points.* A radio tuned to Krebsbach's station was set up on a kitchen table (the humble origin of the table was disguised by a generous use of bunting draped around it) at a school carnival in the area. One Scandinavian farmer, upon being informed that the music emanating from the radio was coming through the air from Vida, declared emphatically that "SUCH A TING YUST CAN'T BE DONE" and explained the whole phenomenon away as being accomplished with the use of a phonograph concealed under the table's bunting.

Early operation of the station was extremely informal, and the financial rewards small. Krebsbach once accepted a rooster as fee for advertising a sale of roosters. In the absence of telephones, farmers often used it to communicate with their homes.

Picture in your mind's eye, if you will, a distinguished silver-haired United States Senator with a white-vested expanse of abdomen of senatorial proportions, clutching the telephone mike in a well-manicured paw, and in all seriousness, campaigning for reelection over KGCX. Some weeks later, his challenger also spoke over the station, but the strength of the signal on that occasion was weak, due to a loose screw in the station's antenna system (which was all of forty feet high) and unjustified complaints of bias on the part of station management were heard.

Vida was by far the smallest town in the U.S. that had a standard broadcast station. Radio maps of the United States issued by RCA about 1927 show a vast expanse of Western territory devoid of dots representing broadcast stations except for the one designating KGCX.

In addition to lacking electric power lines, indoor plumbing (to this day it is still equipped with what the flowery French call *"chalets de necessite"*), and a number of other metropolitan conveniences, Vida also suffered from a lack of public transportation, inasmuch as it wasn't served by any railroad or bus line.

In other words, you just couldn't get there.

Unless perhaps you could hire someone in Wolf Point, Montana, the nearest railhead, to make the trip to Vida in

E.E.Krebsbach is pictured operating KGCX's first transmitter in 1926 at Vida MT.

an open touring car, but in the spring, summer and fall seasons the road was often a sea of mud, and in the winter it was apt to be "drifted in" (a local colloquialism meaning the road is blocked with snow drifts) with the temperature standing at thirty below zero.

Vida's inaccesibility caused visits of the District Radio Inspector (whose home base was Seattle) to be rare, although he was supposed to make an inspection of the station every six months. About 1928, however, Krebsbach received a telegram requesting that he meet the Inspector's train at Wolf Point the following morning at 9 A.M. Hurried long-distance telephone negotiations by Krebsbach arranged for a First Class operator to arrive at Wolf Point on an 8 A.M. train and the trio (Owner, Engineer, and Government Inspector) drove out together to the First State Bank of Vida, where the Inspector retired to another room with his equipment to make certain tests after leaving instructions to turn on the transmitter and to transmit a musical selection. He heard plenty of carrier wave, but no modulation whatsoever. Understandably flustered by the presence of the Inspector, Krebsbach had forgotten to remove the receiver hanging from the converted telephone-microphone, which short-circuited the mike and prevented any music from being broadcast. This difficulty was corrected, only to have the RI discover that KGCX was regrettably occupying a position on the radio dial twenty-two kilocycles away from the spot assigned to it by the Department of Commerce in far-off Washington, D.C. Since contemporary broadcasting stations are now required to maintain their frequency within twenty *cycles*, this was an error roughly equivalent to a London-bound jet landing in Capetown through an error in navigation. The Inspector's last words before boarding his return train to Seattle were, "I set your frequency – now don't fiddle with it."

In 1929, the station received permission to move to Wolf Point (population 2,500) contingent upon increasing its power to 250W. Mr. Krebsbach mail-ordered the required parts for the new transmitter and as a promotion stunt, displayed the various components of the new transmitter-to-be in the show windows of Wolf Point's stores. Even the local Chinese restaurant got into the act; the big plate transformer made a fine prop against which to lean the flyspecked menu.

Bert Hooper, Chief Engineer of CKCK in Regina, Canada, was summoned to Wolf Point to assemble the new transmitter which he did in three days and three sleepless nights.

It looked it.

Hooper traveled to Wolf Point by means of an open cockpit Tiger Moth airplane, leaving Regina on a Sunday, which was a windy, gusty day so typical of the upper Midwest. It was necessary to have a man hanging on each lower wing of the aircraft to steady it until it gained speed for takeoff. Navigation on the flight consisted of flying low enough to read the station names on the prairie railroad depots. Upon reaching the International Border a stop was, of course, mandatory to comply with the requirements of Customs and Immigration laws. However, the actual Boundary itself does not, in the subject area, consist of an impenetrable fence and a gate with barber-pole stripes as depicted in the movies. Rather, it consists of nothing but open and endless prairie land, marked every half mile with a concrete post and with Russian thistle "tumbling weeds" blowing back and forth from one country to the other in utter disregard of protocol. In those days, the Border officials were not stationed at the Boundary itself but rather at the nearest town to the border, which in this case was Scobey, Montana, some sixteen miles south of the International Line. Since Scobey, or for that matter, no other town in the area had an airport, a handy cow pasture was used as a landing field. In landing, the plane narrowly missed several holes excavated by gophers and arrangements were made to avoid such a stop on the return flight by an agreement with the border officials that flying twice around the Scobey water tower would constitute sufficient identification.

ONCE around would have been more than enough, since unidentified airplanes were sighted in that area and era only slightly oftener than Halley's Comet.

While Scobey had no airport, it did have two airplanes, owned by nearby farmers who agreed to escort the Canadian plane to a safe (?) landing field in Wolf Point. Now let Mr. Hooper continue the story as related in a recent letter to the author: "There were two Jennys (American and Canadian World War I training planes which two farmers had bought in Minneapolis for $50 each) that took off with us from Scobey to guide us to Wolf Point. One pilot wore a brilliant red silk shirt and the other pilot a brilliant yellow one. They flew alongside us on the left side. With the up and down air currents, one minute we were looking down at them in their cockpits and the next moment they would be above us. Those Jennys were equipped with OX 5 water-cooled engines and were a bit slower than us so we had to throttle back. It was a wonderful sight . . . We collected the various components which Ed (Krebsbach) had placed in the various store windows for advertising, and I went to work. The pilot, Ted Holmes, was anxious to get back to Regina, so I worked around the clock until nearly daylight, around 3:15 A.M. Wednesday morning. Just missed a herd of cows in taking off, and to avoid them (we didn't see them until it was almost too late), Ted pulled the stick back into his stomach and we missed the power lines by an inch. We got back to Regina about 6 A.M. The Moth was equipped with an inverted six cylinder air-cooled engine of the latest English type and we could do all of 90 mph. I still think of those farmers and admire them for their nerve in buying two surplus training planes for $50 and flying them to their farms in Montana from Minneapolis. No licenses were required in those days. Must admit I did a lousy job in building the transmitter, but if I could have had a night's sleep and more time, I could have done a much better job. The plane was needed back in Canada to fly somewhere to pick up a sick person and fly him to a Regina hospital." (Author's note: Mr. Hooper did an *excellent* job, considering the very short time he had available to build a complete broadcasting station. The wonder is that he could do it at all in three days. The transmitters he built for

CKCK were absolutely beautiful in appearance and performance. The $50 purchase price of the war-surplus airplanes referred to included three flying lessons. "Graduation ceremonies" from the miniscule flying course consisted of a handshake from the instructor who then pointed in a generally west-by-northwest direction and said "Montana is thataway.")

It will be related later in this story what excitement was caused in the towns of the area by an airplane circling overhead. It is easy, therefore, to imagine the carnival atmosphere instantly created in Wolf Point that hot and dusty Sunday afternoon when not only a single airplane, but *three* of them started circling the town with obvious intent to land.

Incidentally, Hooper had passed his First Class radiotelegraph operator's examination at age 15 while he was living in Vancouver, B.C. and employed there as a telegraph messenger boy. Shortly afterward, he arrived home for supper one evening to find instructions awaiting him to report as Radio Officer aboard a vessel due to sail at 8 o'clock that evening. An obliging tailor worked overtime to cut a foot or so off the trousers and sleeves of a Radio Officer's uniform to make it fit young Hooper, who was slight of build, and gold thunderbolts affixed to the telegraph company's cap completed his uniform. He rode his bicycle to the dock, arriving at 7:45 P.M., carried the bicycle aboard and in his haste, reported to the Captain with his right trouser leg still furled around his ankle and held in place with a bicycle clip, and became perhaps the youngest "Sparks" ever to sail the seas.

The new transmitter tried hard to look like the latest product of RCA, but (reflecting its hasty construction) the whole effect was touchingly like that of a small boy's drawing of his conception of what a transmitter looks like.

It worked fine.

A tragic note in the history of the station occurred when an operator was electrocuted when he tried to replace a filter condenser in the high voltage supply without interrupting the program.

One of KGCX's more dramatic services to its listeners was to warn ranchers of the area when the nearby and treacherous Missouri River was on one of its rampages. Ol' Big Muddy has a nasty habit of changing its channel without warning, and in the spring, it often piles up chunks of ice, some as large as an automobile, with resultant flooding behind the ice jam.

Basketball was, and still is, a leading divertissement in the small towns of the upper Midwest during the long bleak winters, which hardly ever last more than seven or eight months. (An old cliche in the region has it that the climate of the area can be described as being eleven months of winter, one month of poor sledding, and thirteen months of wind.) Once, when Wolf Point High School was playing Poplar, Montana, the gymnasium was filled to absolute capacity by an enthusiastic crowd. KGCX's announcer, unable to gain entrance, had to borrow a step-ladder and peer through a transom to watch the game and broadcast the play-by-play account. When Wolf Point reached the finals in the state basketball tournament, local pride was outrageous and coverage of the final game was demanded of and provided by KGCX. A telephone was set up on the playing court at Great Falls, and a reporter relayed an account of the game to an announcer in Wolf Point who broadcast the game. One ardent Wolf Point fan, unwilling to wait for the broadcast version of the game, slipped unnoticed into the KGCX studio, and was monitoring the reporter's account of the game on one of the station's extension telephones. When he heard the final score he shouted "WE WON" so loudly it blasted the transmitter right off the air. It was an hour before repairs could be made and the station could give its listeners the final story of the game.

One Saturday evening when the Krebsbachs were out of town (when the cat is away, etc.), an out-of-state dance orchestra (Duke Snyder and His Happiness Boys) which was barnstorming the area, dropped in at KGCX to provide some live entertainment. With the aid of a large supply of Montana moonshine, the Boys lived up to their name, and along with the technician operating the station, became progressively Happier as the evening wore on, and stayed on the air all night, a distinct departure from normal station policy. The hilarity knew no bounds, and as the percentage of alcohol in the blood of the musicians increased, so did the tempo of the music. The human nose can detect and classify over 16,000 odors. Whether alcoholics can smell the fusel oil in liquor from afar or whether they are unerringly guided to a source of free liquor as surely as the honey bee is directed toward the nectar of the flowers by some not-yet-understood sixth sense, is as yet unknown, but in any event, several of the town lushes soon increased the size of the party. Large platter of fried oysters, fetched from the nearby Hotel Sherman, supplied the calorie energy to keep the festivities going at full blast until they were reluctantly stopped on Sunday morning with fond farewells (performed as only a drunk can) on the part of all concerned, *only* because of the necessity for the station's engineer to run in a broken-field stagger that would have delighted the late Knute Rockne with the station's only microphone to one of the town's churches to cover a regularly scheduled Sunday morning church services broadcast. Some of the local merrymakers, stoned beyond ambulation, remained draped over the studio chairs and leered glassily at all comers, while the orchestra retired to its swaybacked bus, which had a particularly apt-to-break-down-any-moment look about it, to sleep it off in time for their evening performance.

In Wolf Point, the station was associated with an oil company and located in that firm's service station. The clatter of a tire iron, unfortunately dropped on the concrete floor when the mike was open, would occasionally assail the ears of KGCX's listeners. Nevertheless, the station continued to fill a need in the area and in 1936 was granted a power increase to 1,000 watts.

KGCX dominated the air in its domain in more than one way during the late Twenties, since the oil company maintained (as a public relations device) an airplane which roamed over an area of a couple of hundred miles in all directions from Wolf Point and which had the box-car size letters "KGCX" painted on the underside of the lower wing. As previously intimated in this story, the appearance of ANY aircraft in the area's sky was cause for great excitement and if it landed, the pilot NEVER had to walk into town, since shortly after landing he and the airplane would be surrounded by a crowd of the townspeople, who had driven out for a closer look at the flying machine and the Superman who could actually make it fly. The pilot, a somewhat taciturn individual who was reported to have laughed aloud once in 1919 or thereabouts and who was cross-eyed beyond any hope of a surgical correction, occasionally cracked up the plane upon taking off or landing and such occurrence was, of course, THE thrill of the summer for the village, surpassing in excitement and over-the-backyard fence conversations even such great events as the of the Yankee-Robinson Railroad Circus With 500 People And 200 Animals, or the erection of the tent for the annual Chautauqua. The writer remembers attending something called a "rally" with his father at the site of an oil(?) well near Mohall, North Dakota, at which A. C. Townley, a Huey P. Long in microcosm, was the attraction and who arrive; via airplane, which "buzzed" the crowd for dramatic effect before landing. To be able to shake hands with Townley was, for many a farmer in attendance, worth a trip ten times the distance. Upon being queried as to whether he was successful in shaking hands with the Great Person, one farmer was heard to reply, "No, but I TOUCHED his sleeve."

Doughnuts and coffee, the latter conveniently heated with the bona fide natural gas genuinely issuing from the well, were served following which samples of oil were extracted from the well (in tomato cans lowered on a string) to further on-the-spot sales of oil stock by the promoters. The great advantage of the well in question was that it produced oil already refined. A digression to include the Townley story was made only to assure the reader that had it been advertised merely that an airplane would be on hand, with rides going for a dollar a throw (without any celebrity being present), a large crowd, albeit not as huge, would still have attended. Thus the choice of an airplane as an advertising gimmick for both the oil company and KGCX was a brilliant one.

KGCX's antenna actually spanned the main street of Wolf Point, with the transmitter-studio-service station building and one antenna tower on the north side of the street and the other windmill type tower situated in a park-like setting on the south side of the street. This arrangement caused the maximum signal to be radiated in an easterly and westerly direction, directly at the most populated areas of the world. Too, the "flat-top" antenna (a type no longer used these days) was very conducive to the production of sky-wave signals which might be reflected back to earth from the Kennelly–Heaviside layer of the ionosphere thousands and thousands of miles from the little Montana town in which they were generated. In those embryonic days of radio, there were few stations on the air and consequently little or no interference between stations. Therefore, once the signal was started on its way from Wolf Point, there was nothing to stop it and it might fall back to earth in Tokyo or Tangiers or Tampa, depending on "skip" conditions and Mrs. Krebsbach mailed out many a printed Official Confirmation postcard from KGCX upon frantic requests from DX listener hobbyists living thousands of miles distant from Montana. It takes but little projection to place oneself in their shoes and imagine their fascination with such an exotic-sounding name as *Wolf Point, Montana*, bringing visions of timber wolves, Indians astride Pinto ponies attacking th? military outpost, etc., to the mind of a listener, for example, in New York City who had never been west of Philadelphia. Remember, this occurred in the area in question.

The unusual placement of the antenna towers caused hardship for a KGCX relief operator who arrived in Wolf Point via evening train at the height of a Montana blizzard. (It is useless to try to describe to those readers who have spent their lives in Southern climes what carnival Mother

Nature engages in during these prairie storms, which usually claim the lives of several persons each winter. Suffice it to say that visibility is often cut to five feet or perhaps less.) This engineer, now an official of the Federal Communications Commission, located the "wrong" tower (the one in the park) and spent a freezing and fruitless half hour in the storm searching for the nerve center of the station. The possibility that KGCX was literally "working both sides of the street" never occurred to him. With all stores closed (what manner of customer would venture out in such weather?) no one was around to help the lost engineer and it was only by sheer accident that he stumbled into the other tower across the street and into the safety of the transmitter room.

Sparse population in the Wolf Point area, coupled with adverse business there, precipated a financial crisis for the station in the early Forties. With total income for the year 1941 down to only five hundred dollars, it was obvious that *something* had to be done, and quickly. Eyeing a map of the area, Mr. Krebsbach wisely decided that a move to Sidney, Montana (population 5,000) with establishment of auxiliary studios in Williston, North Dakota (population 10,000) was the solution, and the Federal Communications Commission agreed. The move was accomplished in 1942. A complication arose when a major piece of the equipment which had always performed beautifully in Wolf Point, stubbornly and positively refused to work at all in Sidney.

In 1948 KGCX secured another power increase which required the erection of a second tower to protect another station hundreds of miles away from interference by KGCX. A consulting engineer was imported to Sidney to properly locate the new auxiliary tower. After much computation with his slide-rule, he selected the exact spot and the big job of erecting the two-hundred feet high structure was completed. When the last turnbuckle was tightened on the last guying cable, he dusted off his hands and stepped back with a smile of satisfaction to admire his work. As a final-final check, he again consulted his slip-stick. Gradually his eyes opened wider and wider until they approached the size of Satsuma plums, realizing that he had made a monumental error with the result that the whole tower had to be taken down piecemeal and re-erected at a spot some twelve feet away.

During the Fifties, KGCX employed an announcer who, like the late Jack London's character in *Burning Daylight*, attempted to burn the candle at both ends by adopting a daily schedule of announcing all day and doing all night what comes naturally to young men. While the spirit was willing indeed, the flesh sometimes revolted from lack of rest, and the strains of Debussy's *Clair de Lune* or other soothing music from the studio speaker occasionally dropped the fledgling announcer off into the kind of sleep bordering on that from which there is no awakening, with the result that KGCX's slot on the radio dial became deathly and apparently permanently silent. On an automobile trip a hundred miles or so distant from Sidney, with his car radio tuned to his favorite station, Mr. Krebsbach noticed this state of affairs and (apparently considering it to be somewhat less than ideal) made haste to the nearest town and the nearest telephone to correct the situation. The thoughs of Mr. Krebsbach during the comatose announcer incident are not known to this writer, but I am sure his fine Christian faith remained unshaken, in contrast perhaps with that of The Poor Sisters of Perpetual Adoration, who were recently bilked of nearly two million dollars in a Texas oil scheme and who may now be possbily at least a nickle's worth less perpetually adoring.

The original KGCX transmitter has been lost in the shuffle from town to town. Still preserved, however, is the 32V to 1,000V motor-generator (which provided the high voltage for the type 210 final tube) its armature frozen fast, never again to help waft the music of the A&P Gypsies into the blue Montana skies.

The $125 investment in the original transmitter has paid off handsomely; this writer estimates that KGCX is currently worth at least a third of a million dollars and perhaps more.

With comic incidents long past, modern equipment operating dependably, and with financial problems solved, the seven and one-half watt baby has grown into a powerful five thousand watt giant which blankets a large area with its Mutual network and local programs.

Fortune has smiled twice upon Wolf Point since, for the second time, it had a radio station thrust upon it when, in 1957, after business conditions had improved there, Mr. Krebsbach returned to his old stamping grounds after an absence of fifteen years and built KVCK there, a one-thousand watt sister of KGCX.

Mr. Krebsbach, although in declining health, continues as this is being written as owner and General Manager of KGCX and deserves the heartiest of congratulations for building his station from a back-room hobby into the respected and important entity it is today.

Nikola Tesla—
Master of Electrical Energy

J. L. Elkhorne

"Give me a lever long enough, and a fulcrum on which to rest it," Archimedes said, "and I will move the Earth." Nikola Tesla, a nearly forgotten man from a forgotten country, who was perhaps the greatest scientific genius in human history, needed no such prerequisites as Archimedes. Tesla succeeded in resonating the globe as though it were only a child's toy.

This man, who dared presume his power as significant as a god's, had humble beginnings. As the night of July 9, 1856 was drawing to its end, Nikola was born, the second son of a Serbian Orthodox clergyman in Smiljan, Lika, Croatia. (Croatia is now one of the six republics of Yugoslavia.) Djouka Tesla, Nikola's mother, was unable to read or write, yet had an excellent memory and literary facility. She could recite long passages from the Bible and thousands of verses of Serbian national poetry and was an accomplished seamstress. Nikola's father, Milutin, had had an excellent education, though, and had started a career in the military, only to enter the ministry shortly after he married.

Tesla forebears had given a long line of sons either to the military or the Church, and Tesla's father intended Nikola to become a minister. The elder son, Dane, had a brilliant mind and the family expected him to bring them honor as a sicentist or engineer. But Dane died when he was twelve years old, as a result of an accidental fall. It has been speculated by some writers that Nikola had to succeed in his brother's stead, perhaps because he felt guilty. This author thinks rather that Tesla's choice of vocation was the inevitable result of natural ability.

Young Nikola had an inherent mechanical aptitude which evidenced itself many times in his early years. When he was only four years old, he built a water wheel for the creek which flowed near his home. During his childhood he made many curious devices. One such was a popgun which fired a ball of wet hemp. He built and sold these to his fellow children until a rash of broken windows ended this enterprise. He turned to archery and went from bow-and-arrow to crossbow and then arbalest. He was fascinated by the idea of flying and when he was twelve years old, made an unsuccessful parachute jump from the barn, using an umbrella.

He evidenced an original turn of mind from his earliest days. Once he noticed that after lightning struck from the dark masses of clouds that torrents of rain would fall, and decided that one might be able to control rainfall by controlling the lightning. Years later he had succeeded in producing artificial lightning, but was never able to convince the U.S. Patent Office that weather control might be possible.

In school his talents quickly led him to the head of his class, particularly in mathematics, a subject he favored. He detested resorting to paper and pen when it was so easy for him to solve any problem in his mind almost as quickly as the teacher had given it. This ability led to his being suspected of cheating and he was not given a passing mark. Nikola went to the director of the school and demanded an examination. The test was duly given, and in the presence of the director and the teacher, Tesla solved problems far beyond his years. Suspecting that he had somehow gotten the answers to the standard examination, the officials departed from it to throw even more difficult problems at him, which he solved equally well. The astonished men could only acknowledge that Nikola Tesla possessed an astounding ability and passed him.

At fifteen Tesla continued his education at the Higher Real Gymnasium in Karlovac, Croatia, which corresponds to our college-level training. He completed the four year course in three years. During this time he was living with an aunt and her retired army officer husband. The woman felt that Nikola was of delicate health and that heavy meals would have a bad effect on him. Tesla remembered this as the hungriest period of his life. While he was at Karlovac, he took many hikes along snow-covered mountain trails. One day he began rolling snowballs down a slope, trying to see how big he could get one. He succeeded too well and sparked an avalanche which roared down the mountainside and diverted itself harmlessly in a field, only just missing some farm buildings. Tesla was horrified at the near disaster he had created, but wise enough to recongize that the tremendous power of Nature could be harnessed by relatively small applications of man's power.

When he had completed his studies at the Gymnasium, his father had written urging him to take a hunting trip rather than returning home. Tesla impulsively disregarded his father's advice and went home to find the area in the grip of a cholera epidemic. Worse than this was the elder Tesla's desire to see his son study for the ministry. Milutin Tesla was not wholly autocratic, however. He knew that if his son did not enter the church, he would be required to serve three years in the army. Nikola felt he was caught between Scylla and Charybdis, about to be crushed by alternatives he could not stand. Three years as a soldier seemed to him three years as an unthinking robot, compelled to the mindless disciplines of drill and routine. And a life in the ministry would leave him no time to learn Nature's secrets. Nikola had decided that he wanted to be an engineer.

The three years of undernourishment and the spiritual anguish he faced now so weakened him that he succumbed to cholera. For months he lay ill, one sinking spell leading to another. The family doctor finally announced that he could do nothing more for the boy and the family should prepare themselves for his imminent death.

Now the elder Tesla was facing his own crisis. One son had already died. He had pledged Nikola to the church, but if the boy died, the pledge would be unfulfilled. In anguish the father asked the dying son what he could do. Nikola whispered that he could get well if only he could study engineering, and the despairing father quickly agreed. The deathbed crisis passed, Nikola's will strengthened, and he began to recover. Tesla wrote in later years that no magical event had taken place, but rather his recovery had been because of distasteful but potent medicine his mother prepared.

Subsequently the army declared the convalescing Nikola unfit for service and he was now free to pursue his goals. The elder Tesla's influence with other members of the family in the military for this decision has been suggested, but is not definitely known. It is a fact that the father sent Nikola away for a year while the decision was being rendered. During this year Nikola amused himself with such fanciful projects as a proposed subterranean tunnel between Europe and America through which containers could be propelled by water pressure. Tesla quickly realized that drag would make the project unworkable, but nevertheless enjoyed such pursuits as a stimulus to his imagaination.

In 1875, nineteen-year-old Tesla entered the Polytechnic Institute at Graz, Austria. In his effort to prove himself worthy of his father's reconsidered decision, he took twice the normal amount of subjects and limited himself to less than four hours' sleep a night. At the end of the year, he returned home with the highest possible marks, fully expecting his father's praise. Instead, his father disregarded the effort his son had made and critized him for endangering his health. It was only years later that Tesla discovered that the dean of the technical faculty had written Milutin Tesla to say that Nikola was "a star of the first rank but will kill himself from overwork."

In his second year at the Institute, Tesla limited his studies to physics, mechanics and mathematics. During a demonstration of a Gramme motor/dynamo, Tesla remarked that the commutator sparking indicated power loss. His instructor, Professor Poeschl, patiently explained that the commutator was necessary to provide a direct current output. Tesla responded that by discarding the inefficient commutator and using the natural alternating current from the rotating armature, far greater efficiency could be attained. A hail of invectives greeted Tesla's conjectures, comparing them with such foolishness as perpetual motion machines. Since early experimental motors would not run, it was assumed that the positive and negative cycles were cancelling each other. Some ac generators were used about this time, but only for resistive loads, such as street lamps. Everyone knew you could never get a motor to work on ac!

Though Tesla bowed before the authority of his professor, he could not get the concept out of his mind. Plan after plan was imagined and discarded. Tesla had had, from his earliest childhood memory, an amazing ability to visualize objects. When he thought of something, it appeared before him, as real to him as any object in the external world. It took some time for the child to realize that other persons could not perceive these visualizations of his. The adult Tesla always said that he perfected his models in his mind, and that there they were so real that he could see signs of wear and, in the case of rotating machinery, could actually tell whether or not it might be out of balance. His visualizations were accurate to the thousandth of an inch; the piece parts he required did not need any trial-and-error machining. Years later, in America, the staff he had in his laboratory had a hard time keeping up with him. If his precise verbal instructions did not seem clear to a workman, Tesla would make a small, neat sketch on the handiest piece of paper. It is said that no matter what size paper Tesla picked up, his sketch was sure to be no more than one inch across, yet perfectly detailed.

This uncanny visualization, coupled with his amazing mathematical ability, certainly aided his inventive efficacy. Tesla had what we would call a photographic memory, and was able to quote – from beginning to end – Goethe's *Faust*, as well as a good deal of Shakespeare, and other classics. He committed the logarithm tables to memory, so that he would not have to waste computation time in reference.

Since he had heeded his father's advice for his second year at the Polytechnic Institute and had not taken as many classes, he rounded out his activities with billiards, chess – and poker. As he could visualize many moves ahead on the chessboard, it was only a short time before he had conquered all his fellow students. He then set about organizing a chess team which challenged other schools – the first known example of intercollegiate activities. His first poker game was a memorable one – the companion who took him to the game had promised a lamb for the shearing. By the end of the evening, the lamb had won everything. Then he amazed everyone by returning, to the cent, what each player had lost. For Tesla it was only a means of relaxation, not a profit-making venture. Many times he returned to the card table, and it was the same. Then one night luck or his own ability let him down. He calmly bet the tuition money had had received for the next year's study, and it was quickly gone. The game was over – but no one offered to return the money as Tesla had. This was a painful lesson about the nature of men, and Tesla was deeply shamed that he had lost his parents' careful savings. Yet he could do only one honorable thing, and that was to confess all. He returned home, found his mother, and told her what had happened. Djouka Tesla sagely understood. Where his father would have scolded him for immoral activities, the mother knew that her son was obsessed. She gave him what little remained of their savings, telling him that he had yet to learn a lesson. When Tesla returned to the card sharps, they expected only more loot. Instead, Tesla, playing with steel determination, won everything. When the game was finished, the players expected their money to be returned. This time, Tesla kept it. He had regained his tuition, and the money his mother had advanced him was returned. He made a solemn oath never to play cards again.

When he had completed his studies at the Institute, Tesla took a job at nearby Maribor with a tool-and-die works which was manufacturing electrical equipment. The money he saved enabled him to study a year at the University of Prague. Upon graduation, he traveled to Budapest, where a telephone central office was being built. The excellently-educated engineering graduate found that no responsible engineering position was open to him. Ironically, the job he was able to accept was that of draftsman with the Hungarian Government Telegraph Office. Some forty years later, he wrote that it was "at a salary which I deem it my privilege not to disclose!" He also recalled that, "By an irony of fate my first employment was as a draughtsman. I hated drawing; it was for me the very worst of annoyances." Yet Nikola Tesla could not do a job poorly, and it was not long before his ability was noticed and he was promoted to more responsible work, and finally made chief electrician of the telephone company. At only twenty-five years of age, he found himself engineer-in-charge of an entire system.

It was to his liking. He found himself fully occupied during the day, limiting himself to a five-hour rest period, in which he tried to keep up with current technical journals, sleeping only two hours a night. At this time he invented the first "speakerphone," a loudspeaker apparatus with which a roomful of people could hear a telephone conversation. Tesla never bothered to patent it, though the company put it into service. Thirty years later he remarked that it compared favorably with then-current loudspeaker designs.

All the while his mind was constantly working on the concept of alternating current, but he had not yet solved the basic problem. Then the long hours of toil, welcome though they were to him, took effect and Tesla had a breakdown. Chief symptom of the unique nervous disorder he developed was a highly abnormal sensitivity. Tesla wrote: "I could hear the ticking of a watch with three rooms between me and the timepiece. A fly alighting on a table in the room would case a dull thud in my ear . . . in the dark I had the sense of a bat and could detect the present of an object twelve feet away by a peculiar creepy sensation on the forehead." Doctors pronounced the malady incurable, though they did not know what it was. As quickly as it had come, it went. While he was convalescing, his assistant, Szigeti, went with him for a walk in the park one afternoon. Tesla was pleased that the illness had not affected hisı memory and, looking at the sunset, began an appropriate quote from Goethe. Suddenly, before him, was the alternating-current apparatus he had thought about for so long. "Watch me reverse it!" he cried, throwing an imaginary switch. Szigeti feared that his boss had gone 'round the bend. Tesla calmed himself enough to explain. Evidently his mind had unconsciously completed the assigned task while he was ill. Tesla picked up a stick and sketched the circuit diagrams on the dirt path, explaining each detail to Szigeti. Tesla would use a two-phase ac in the field coils of his motor, which would produce a rotating magnetic field. This magnetic whirlwind would – by induction – pull the armature around with it. The armature would have wound coils closed on themselves, requiring no external electrical connection. Thus, in a blinding flash of inspiration, Tesla had grasped the answer. There would be no commutator to cause power loss; the only wear point of the motor would be the mechanical bearings.

Years later, a president of the American Institute of Electrical Engineering said: "The work of Nikola Tesla in his great conception of his rotary field seems to me one of the greatest feats of imagination which has ever been attained by the human mind."

The rotary magnetic field was more than an invention. It was a basic discovery, and it was this ability to perceive new insights into nature's wonders that set Tesla apart from other men. While those such as Edison made improvements on existing theory, Tesla forged ahead into an uncharted wilderness, where man had not gone before. It was Tesla's gift that he could discover – but the gift contained tragedy, in that other men could not, or would not, follow.

It was in February, 1882, that Tesla grasped the answer to the alternating-current problem. It would be six years before he would convince men that this was a revolutionary, wonderful new system. His immediate problem was more mundane – the job he had held was ended, as his employer had sold the business. Puskas, the employer, wrote a letter of recommendation for Tesla, which enabled him to get a job with the Continental Edison Company in Paris.

Tesla felt that the cosmopolitan city of Paris would give him the opportunity he needed and he swiftly took the job. Even more swiftly, he found that his employers were not interested in any "crackbrained schemes." Tesla tried to interest businessmen in his alternating-current system. Friends warned him that someone would steal his discovery, but that was hardly the case. Tesla could not even give it away.

In the meantime, his position with Con. Edison had been done in the usual flawless manner. Tesla was assigned as a roving trouble-shooter, and in 1883 was sent to Strasbourg, Alsace, then part of Germany, to straighten out an embarrassing situation. A new power plant had been built and Emperor William I was present at the dedication ceremonies. Unfortunately, the plant had been wired by men who were less than adept at their job, and when the master switch was thrown to activate the new lighting system, a short circuit blew out one wall, showering bricks and debris on the dignitaries. The Germans understandably would not accept the plant in its present condition and Tesla was dispatched to the rescue.

He soon put matters right, even in the face of German bureaucratic bungling. He writes of the "efficiency" of a simple matter of placing a hall light, in which the whole chain of command had to be consulted before the light could be installed at the very spot Tesla had suggested to begin with. When he was not taking time in matters of this sort, however, he was free to work on his own. He had brought some materials with him, and rented the facilities of a machine shop near the railroad station for evening work. There he built his first induction motor. Because he could visualize the final design so accurately, he did not need blueprints. He was a fussy worker, however, and the precise machining and polishing of the parts took time. The individual parts did not need cut-and-fit partial assembly, but went together into a finished product the first time.

It was a dramatic moment. Tesla had thought the idea through in his mind, but it was new, and it had never been tried before. Perhaps he had deluded himself, maybe it was only wishful thinking. He could only find out one way – he threw the switch. The motor hummed, the armature turned. Almost instantly, it had built up full speed. Tesla threw the reversing switch. The armature stopped, began to revolve in the opposite direction. It was clearly a success.

Tesla had made many friends in Strasbourg, impressed many people because of his efficiency and knowledge. Now he went to the mayor and various businessmen. Perhaps the sight of a working model would suffice where words had failed. But it was the same. No one seemed interested in what was the most significant commercial development in electrical engineering.

The despairing inventor returned to Paris, where another disappointment awaited him. He had been promised a substantial bonus on successful completion of the Strasbourg project. He had delivered. Now, the managers of the works sent him on a wild-goose-chase. The treasurer would pay him, only the treasurer said he did not have the authority to issue a draft; the operations manager understood, but he had not authorized the trip, etc. etc. Tesla gave his resignation in disgust.

Charles Batchellor, one of the company administrators who had been friendly with Tesla, advised him that he would have a better chance of achieving success in America. Batchellor was a close personal friend of Thomas Edison and gave Tesla a letter of introduction.

Once the idea was in Tesla's mind, it didn't take long for him to settle his affairs. He sold his possessions and books and bought a steamship ticket. On the way to the station where he would catch a train for the seaport, he was robbed. He managed to get to the docks but the steamship officials would not let him board without a ticket. He persuaded them that if no one showed up before departure to claim the reservation, they should accept his story as true. He arrived in America with four cents in his pockets, a book of his own poems, a couple of technical articles, some notes on a mathematical problem and on the design of a flying machine – and a wealth of inventive genius in his mind.

He presented himself to electrical wizard Edison quickly and angered the great man immediately by telling him of his alternating-current system. Yet Edison was quick to recognize the ability of the young immigrant and impressed by the letter Batchellor had written. Never one to ignore the knock of opportunity, Edison hired the well-educated and experienced engineer for eighteen dollars a week – hardly more than he paid the average mechanic in the shops.

Tesla was impressed on their initial meeting by Edison's forceful personality and recognized that the practical man had done quite a lot with no formal training. He wondered if his own educational process had been a waste of time. He was quick to learn, however, that it had not. Edison firmly believed in the trial-and-error approach to all things, and much later Tesla wrote that "a little theory and calculation would have saved him ninety percent of his labour."

Tesla was initially assigned minor, routine work, but when he showed his talent and his dedication by working eighteen hours a day, Edison soon trusted him with important tasks. One of the most significant concerned the steamship *Oregon*. Edison generating equipment had been installed on this most up-to-date liner of the day and worked well for many months but eventually broke down. Edison sent his lesser lieutenants to make repairs, but they failed. The *Oregon* did not sail as scheduled, and Edison was faced with financial and personal embarrassment because of his equipment. In desperation he sent hard-working Tesla to try his luck. Working through one night with the assistance of a willing crew, Tesla rewound the armatures which had short-circuited. Tesla met Edison and Batchellor, who had returned from Paris. When Tesla reported success, Edison remarked to Batchellor that Tesla was as good as Batchellor had said, and "indeed, he's even as good as he thinks he is!"

But their good relationship was soon at an end. Tesla was already discomfited by Edison's refusal to consider an alternating-current plan. It is true that Edison had made a $2 million investment in New York City on a dc distribution system. Naturally he did not want to see this expenditure threatened.

The final blow came as the result of a statement Edison made to Tesla concerning some research problems. Edison told him it would be worth fifty thousand dollars if he could come up with the answers. In all, Tesla's improvements led to twenty-four new dynamo designs utilizing highly efficient short field-core magnets, some new automatic controllers, and several patents taken out in Edison's name. Tesla had delivered the goods and eagerly awaited the promised bonus. When it was not forthcoming, he asked the great man and was told, "Tesla, you just don't understand our Yankee humor."

It was spring, 1885, and he had been with Edison less than a year, but he could no longer go on. As Con. Edison had cheated him in Europe, so he was led down the garden path in America by Edison himself. Tesla's reputation had developed well, and some promoters approached him with the idea of starting a street and factory lighting company under his name. Tesla offered his alternating current system to them, but they rejected it in favor of quick profits in this utility venture. Tesla agreed to develop a practical arc lamp for shares in the company. For about a year he worked at a very small salary, producing all the equipment the businessmen asked for, and taking out several patents. When the system was well under way, Tesla was given stock shares and was quickly manipulated out of the company. He found that his share certificate was worth very little. The United States was undergoing a depression and Tesla was without income. His former associates, not content with having eased him out of the company, now made spurious claims about his unreliability. And Edison certainly had nothing good to say about him.

Tesla underwent such a period of hardship during the next year that he would never discuss it in later life. It is known that he did occasional electrical repair jobs, when he could find them, and that he worked as a common laborer. During the winter of 1887, Tesla was working as a ditch digger. The crew foreman, a former stock broker who had lost everything in the market, became interested enough in Tesla's theories to introduce him to A. K. Brown of Western Union Telegraph. Brown and an associate financed a laboratory for Tesla, and organized the Tesla Electric Company in April, 1887. Tesla's lab was at 33–35 old South Fifth Avenue (later West Broadway) and, not far from the Edison works. The models Tesla now built were identical to the designs he had conceived five years before. In all this time he had made no notes – the concepts were firmly locked in his memory. By October of that year, Tesla had filed for a patent. He wanted a single patent for the discrete elements which comprised his system. The U.S. Patent Office was horrified at the idea of such a sweeping omnibus approach. They insisted on a breakdown to seven separate sections. By the end of the year he had filed for and received thirty basic patents.

The scientific world which had ignored Tesla for so long now could not contain its amazement. As the import of his system was grasped, he was praised as the scientific genius of the age. The lecture he delivered on invitation to the American Institute of Electrical Engineers on May 16, 1888, is a classic of the electrical engineering field. The theory and practice he presented are the basis of the system we still use today. In one stroke he accomplished an engineering breakthrough of such magnitude that no comparable development has been presented since. Tesla was on the crest of an inventive wave that would last for thirty years.

Fortunately for Tesla – and for the world – the man of commerce who could bring this scientific feat out of the laboratory and into the everyday world of engineering practice approached. This was George Westinghouse, inventor in his own right, and head of his own company. Westinghouse was not committed to direct current as Edison was. In Tesla's system, he saw the revolutionary means of long-distance power transmission that would reap profits such as Edison's plant could only wish for. Westinghouse offered Tesla one million dollars for the rights to the group of patents, which now numbered forty. Tesla agreed, provided that a royalty based on equipment produced was also paid. A conservative estimate was later made that Tesla should have received an additional $12 million in royalty payments.

This dynamic duo was the worst possible threat to the Edison system. What Tesla – the man of vision – lacked in financial awareness, Westinghouse – the successful entrepreneur – could supply. The United States was leaving a phase of depression and entering one of explosive capitalistic growth, in which giant industrial empires were born. Westinghouse, in his eagerness to make the most of the Tesla system, expanded his corporation by consolidating a number of smaller companies and also by attracting much outside capital. The time came, however, when Westinghouse's board of directors overruled him on the matter of royalty payments to Tesla. Even though there was a legal contract in which the payments were stipulated, the board argued to an unwilling Westinghouse that such payment would seriously endanger the stability of the corporation. Further, if Westinghouse insisted on honoring the contract, much of the outside capital would be withdrawn. As an inventor himself, Westinghouse understood the justice of the royalty payment; as a businessman, he could not bear the thought of the empire crumbling. He went to Tesla and explained the situation. Tesla had become a close personal friend of Westinghouse and understood the problem. Further, he was more interested in seeing his system operating on a successful basis than in collecting a payment legally his, which would wreck any chance of that system's development. Tesla tore up the contract and the Westinghouse empire was saved.

Tesla had had to give half of his million dollar initial payment to Brown and his associate, but the half which was left served well enough to maintain the most splendid research establishment seen in America at that time. Tesla embarked on a program of research that would have broken most men. He began his monumental studies into high-frequency phenomena.

The Edison faction had not been inactive during this period and in a calculated campaign to destroy the alternating-current disciples, began a whirlwind of adverse publicity. Edison and his friends gave Sunday afternoon demonstrations on the evils of ac by electrocuting cats and dogs for the edification of visitors.

Edison wrote: "Just as certain as death, Westinghouse will kill a customer within 6 months after he puts in a system of any size. He has got a new thing and it will require a great deal of experimenting to get it working practically. It will never be free from danger."

A propaganda campaign of immense proportions arose. Most of the so-called scientific proof of the horrible dangers of ac erupted from Edison's laboratory in West

Orange. Misleading statements were issued to the press. Pamphlets were distributed, warning the people that it would soon be a matter of taking one's life in his hands to merely walk the streets, constantly at the mercy of the lethal high-tension wires. A further suspicious fact is that a former laboratory assistant at West Orange, H.P. Brown, began lobbying and lecturing for the passage of a bill in the New York state legislature for the provision of death by electrocution.

Such a bill was passed in 1888, and H.P. Brown, now a consultant to the state, authorized the purchase of three Westinghouse alternators to be installed at Sing Sing Prison. George Westinghouse protested this particular use of his equipment, but the authorities pointed out blandly that dc generators could hardly provide the high voltages necessary. On August 6, 1890, convicted murderer William Kemmler was to be put to death in a secret ceremony. The engineers who had installed the electric chair had been perhaps more frightened of it than was necessary. It was reported that the power was insufficient to cause death. Repairs had to be made, and the execution repeated, resulting in "an awful spectacle, much worse than hanging . . ."

Meanwhile, Tesla had not been idle. He had, of course, selected 60-Hertz operation as best for commercial power applications. Indeed, he had opposition from Westinghouse engineers at the Pittsburgh Plant where the new apparatus was being developed. They preferred a 133-Hertz system, partly because of decreased cost of core materials needed at the higher frequency. Tesla left in disgust, even though Westinghouse was able to offer him $24,000 a year salary. Shortly thereafter, the engineers did select our familiar 60-Hertz system as the standard. Tesla remarked that the year he spent in Pittsburgh was wasted in minor design problems, and that he was not free for creative work. In the next four years, he was granted 45 additional patents on polyphase current distribution.

His researches into higher frequencies had led him to the discovery that as the frequency increased, less and less iron was necessary in transformer cores. Utilizing conventional rotary dynamo technique, he built devices which produced up to 10,000 Hertz. The next step was the production of even higher frequencies and the devleopment by him of the air-core high-frequency transformer known and loved by all as "the Tesla coil." Two decades later, F. W. Alexanderson was developing high-frequency ac dynamos for high-power wireless transmitters for the government, along the same lines.

Incidentally, Tesla's choice of 60-Hertz operation made possible cheap electric clocks, driven by synchronous motors, a fact he pointed out freely.

Along with the high frequencies Tesla was now producing, were also high potentials, so much so that conventional insulating methods were ineffectual. It was at this time that Tesla produced the technique of oil immersion, which had great commercial importance. He had soon reached the practical limits of frequency from rotary dynamos. Now he was exploring the field of resonance phenomena. He developed the technique of electrical tuning in 1890, one of the basic principles of radio.

Utilizing Lord Kelvin's theory of the damped oscillation wave of a discharging condenser, Tesla developed means of charging a condenser by low voltage, then using the "disruptive discharge" through the primary of his air-core transformer, deriving very high-frequency oscillations and extremely high potentials in the output. He discovered the heating effect in the human body of the high frequency currents and thereby laid the principles for medical diathermy. Here was another pioneer discovery for which he took no patents nor credit.

He lectured once more before the American Institute of Electrical Engineers on May 20, 1891, on the subject of high frequency currents, and demonstrated a variety of phenomena. With apparatus then, he was able to achieve a spark discharge of over five inches, indicative of a potential in excess of 100,000 volts. The induction coil he used was energized from a generator of his own design working at 20,000 Hertz. He described the various discharge phenomena exhibited under varying conditions of potential and frequency. In his research he had discovered that the nerves of the human body could not react to currents higher than 700 Hertz. He recognized that high voltage as such was not lethal, as he had taken the high-frequency discharges of his larger apparatus many times without ill effect. Partly it is because of "skin effect," in which the high frequency tends to travel across a surface of a conductor. Also the output of Tesla's apparatus had a low current density. He wrote that slight changes of potential, current, and frequency could work together to form a lethal shock

He said that the surest way to electrocute a person was to subject him to sustained direct current, but that the most painful means of killing would be to use a low-frequency alternating current. It is conceivable that a high-power, high-frequency current could kill swiftly and painlessly. But his demonstrations before the Institute were clear proof that alternating currents, as such, were quite harmless. Tesla repeatedly put himself into the high voltage secondary circuit of his coil, in the process of lighting bulbs of his own design, with no discomfort.

Part of the awe-inspiring show was the fact that the forms of illumination Tesla was demonstrating had never been seen before. Instead of the ordinary Edison bulb for light, Tesla was using ordinary wires in normal atmospheric pressure to demonstrate luminosity effects. He had vacuum lamps, lamps with other gases, lamps with extremely high internal pressure – and most of the above had but one terminal. Others had none at all!

The production of a practical means of commercial power and now the development of a whole new spectrum of alternating currents – both within three years – established his reputation firmly with the scientific community. All at once he had more offers for lectures, after-dinner speeches, demonstrations, and consultations than he could cope with. One of the problems was that his high sense of intellectual honesty and originaltiy required him never to duplicate the material in a lecture; thus each appearance of this prodigal genius before the public was a unique event.

He was invited to lecture before scientific bodies in England and on the Continent. In Feburary, 1892, he gave another lecture on high-frequency currents before the Institute of Electrical Engineers in London. Sir James Dewar asked Tesla to repeat the lecture before the Royal Society. Tesla had other plans and demurred, but Sir James took him to Michael Faraday's chair and plied him with the remaining private stock of whisky which had belonged to that earlier genius of electrical invention. After such a singular honor, Tesla could do not less than agree.

While he was touring Europe he received word that his mother was gravely ill. He arrived at her home in time and was able to talk with her that day. She died during the night. The strain of rushing to her deathbed caused a patch of hair on his head to turn white overnight, but a month later it had regained its natural color. His father had died years before, while he was still a student.

When Tesla returned to America he realized that his social success had cut into his research time tremendously and he turned away from all such engagements in the future, preferring to devote himself to useful work in the laboratory. During this period he experimented with such diverse items as high-frequency currents, mechanical oscillators, X-rays, and astronomical studies. He prepared a large exhibit for the 1893 Columbian Exposition, where Westinghouse had won the contract to furnish power and lighting equipment. During frequent demonstrations at this World's Fair, Tesla passed one million volts of high-frequency electricity through his body, to turn copper plates molten or to light special bulbs of his own exotic design. The public was thoroughly convinced that ac was not the danger Edison and his followers claimed.

The final blow the dc faction received was the harnessing of Niagara Falls. The child Tesla, having seen a picture of Niagara and perhaps influenced by that first toy waterwheel, prophesied that he would someday make the Falls work for him. Now that dream was to come true. A charter for developing power at the Falls had been granted in 1886, and in 1890, Edward Dean Adams, head of the

Cataract Construction Company, organized the International Niagara Commission. Lord Kelvin, the famous British scientist, was made chairman of this body, which was to determine the best method. A prize of $3,000 was offered for the best plan submitted. Tesla was frantic to toss his hat in the ring, but George Westinghouse pointed out that the prize would be insufficient payment for value received and persuaded Tesla against entering the competition. A further complication was that Kelvin favored direct current. However, in 1893, when none of the major manufacturing concerns had submitted plans to claim the prize, the commission asked for bids. By now the Edison faction had capitulated and paid for the rights to utilize the ac patents. Westinghouse's bid for the generating plant was accepted on May 6, and the by-now General Electric Company, Edison's empire, was chosen to build the transmission line to Buffalo, twenty-two miles away. After the completion of this project, which was described as the "most tremendous event in all engineering history," Kelvin admitted that alternating current had many more advantages, and further stated that "Tesla has contributed more to electrical science than any man up to his time."

Ironically, during this period of commercial development of the polyphase ac system, Tesla was referred to as an imitator by the British press, which claimed that he had used as the basis of his system the apparatus of a physicist at the University of Turin, Professor Galileo Ferraris. It has been proven that Ferraris first presented a paper on "electrodynamic rotation" in 1888, six years after Tesla's discovery of the rotating magnetic field, and indeed, several months after Tesla's application to the U.S. Patent Office. Ferraris had developed an alternating-current device as a demonstration of circularly polarized light and stated specifically that the principle behind his model could never be developed as a practical power unit. Yet chauvinistic English publications, which had received notice of the Tesla system, ignored it.

In 1891, another pretender to the throne arose, in the person of Dolivo Dobrowolsky, at the Frankfurt Industrial

Exposition of 1891. Dobrowolsky claimed the invention of the first practical ac motors, and later reduced his assertions to a greater efficiency of his three-phase motor than that of the original two-phase Tesla induction motor. However, the chief engineer of the project, C.E.L. Brown, completely smashed Dobrowolsky's claims by writing to the *Electrical World* that "the three-phase current as applied at Frankfurt is due to the labors of Mr. Tesla and will be found clearly specified in his patents."

Part of the problem during this period of massive technological change was that the lines of communication and publication of new inventions were very slow. And there were those that put forth fraudulent claims, so that they could reap the rewards. But at the same time, errors of fact have been made by writers too lazy to do proper research.

Finally and unequivocally, Tesla was granted the credit due him, by scientists, engineers, and editors throughout the world. It is certain that the many persons who saw Tesla's demonstrations at the Columbian Exposition would not soon forget him. By 1900 there were many usurpers and infringers on the Tesla patents. The Westinghouse Co. took about twenty suits to the courts, and the Tesla patents were upheld in every case. Judge Townsend of the U.S. Circuit Court of Connecticut in September, 1900, wrote: "It remained to the genius of Tesla to capture the unruly, unrestrained and hitherto opposing elements in the field of nature and art and to harness them to draw the machines of men . . . he first conceived the idea that alternations might be transformed into power-producing rotations, a whirling field of force.

"What others looked upon as only invincible barriers, impassable currents and contradictory forces he seized, and by harmonizing their directions utilized in practical motors in distant cities the power of Niagara."

After his European and American lecture tours and his triumph at the World's Fair, Tesla withdrew from public and social life. The next two years were full to bursting. Induction coils had led him from the curious phenomena of high-frequency to the realization of the nature of electrical resonance. He planned to develop wireless communication through the earth by means of his air-core transformers. Soon he progressed to the notion that power itself could be transferred. Even working eighteen hours a day, he had not enough time for the ambitious program he undertook. During this period he discovered, independently, X-ray phenomena of the type that Roentgen would soon announce. But Tesla never claimed his system was similar to Roentgen's. All he noted at this time was a type of "very special radiation." With it, he was able to get shadowgraph pictures through a human head – at a distance of 40 feet!

He turned his attention to atmospheric phenomena and declared that the *aurora borealis* was caused by expulsion from the sun of particles of high electrical charge. Professional astronomers laughed at such an idea; after all, they knew that the sun was 93 million miles away. He further convinced them of his eccentricity with a casual statement about the "dozen or so" planets of the solar system. Then he announced his detection of mysterious rays bombarding the earth, of hundreds of millions of volts in energy. A decade later, in 1909, W. H. Pickering announced the probability of a trans-Neptunian planet. In 1930, such a planet was discovered photographically and named Pluto. Robert Millikan announced the existence of cosmic rays in 1926 and thereby won a Nobel Prize in physics.

The above were merely side interests. His main curiosity was in resonance phenomena. Aside from electrical oscillations, Tesla had developed several interesting mechanical resonators, testing their low frequency mechanical vibrations and their effect on the human body and on various materials. Noticing the invigorating effect of the vibrations, he was induced to construct a small massager for the barber of one of his assistants – the ancestor of the same massager the barbers of today use. Another of his vibrating machines consisted of a platform on which one could stand. The platform was connected to a large motor which could be adjusted for different rates of vibration and which created invigorating sensations in the human body. Author Sam Clemens had become a close friend of Tesla and visited the inventor's laboratory many times. Tesla admired Clemens and claimed that one of "Mark Twain's" books had hastened his recovery from a youthful illness. One day, Clemens tried the platform and found it so refreshing that he would not get off, even at the urging of his friend, Tesla. Suddenly he stepped down and demanded petulantly to know "where it was." Smothering a smile, Tesla pointed the way to the rest room, where Clemens hastened. At certain rates, the curious platform had an irresistable laxative effect.

Another of Tesla's mechanical oscillators had more serious ramifications. Tesla had noted that gentle application of power at the natural resonant frequency of a material would set up strong vibrations within it, just as it is possible for an opera singer's voice sustaining one note to sometimes shatter glass. Tesla attached his puny device to a strong vertical steel girder in the laboratory. It seemed improbable that any significant action would occur. He activated the device, which was described later as small enough to slip into one's pocket. It was automatic in operation, in that it would "hunt" for the natural frequency of the substance it was attached to and then lock in step and reinforce the resonant wave. At first, nothing could be noted in laboratory. However, in other parts of Manhattan Island matters were quite different. The strata of sand beneath the surface transmitted the vibrations exceedingly well, whereupon they reflected from the granite layer of bedrock. Since the vibrations were sustained by the mechanical oscillator, the power of the waves constantly grew. Shortly, windows were cracking, plaster falling, furniture moving wildly about. The man-made earthquake began at some distance from the laboratory and slowly moved in toward the source center. By the time the shock wave impinged on Tesla's building, local police were deluged with reports. Officers were sent to the laboratory. When they entered, they found Tesla smashing the oscillator with a sledge hammer. He had realized what must be happening and taken the swiftest action necessary to cease his experiment. The police had been right in believing Tesla to be involved – after all, anyone who could make lightning in the laboratory must be responsible for other strange occurences.

Tesla later calculated that his innocent-appearing device was capable of much greater damage. He never released specifics on its construction, prudently, and never resumed such experiments. Later he claimed that with it he could have destroyed the Brooklyn bridge within an hour, or "could now go over to the Empire State Building and reduce it to wreckage in a very short time." At least one published estimate of the time necessary was fifteen minutes! Lest someone think this an idle boast, recall that soldiers break step when marching across a bridge. Further, the infamous "Galloping Gertie," a bridge across the Tacoma Narrows built in 1940, showed all too clearly the result of destructive vibration. When completed, it was the third longest suspension bridge in the country. Four months after its completion, a wind storm, with gusts to 42 mph, started erratic vibrations in its span. After four hours, waves of thirty feet were passing along. Then it began undulating from side to side. Four more hours saw 600 feet of the center span plunge into the Narrows. The 1000-foot side spans followed. Anyone who has seen the film of this bridge swinging from side to side in a mad dance will acknowledge the effects of internal vibration.

On the night of March 13, 1895, catastrophe of another sort hit the Tesla laboratory. A fire razed his building, destroying several years' work, most of Tesla's awards and mementos, what notes he did keep. Fortunately his prodigious memory kept the loss to a minimum, but almost all his experimental equipment was lost. He opened a new laboratory at 46 E. Houston Avenue in July of that year and went on with his work.

By September of 1897, he had filed and received U.S. Patents 645,576 and 649,621 – which have been called the fundamentals of radio broadcasting. He was now ready to demonstrate for the public his new wonders. Always the showman, he booked Madison Square Garden and put on a demonstration of what we would refer to as a radio-controlled boat. He called his remotely-controlled devices "telautomatons." Thousands of people saw a small craft with antennas which could be ordered about at will. Tesla had had constructed an immense tank, in which the boat was placed for the demonstration. The model craft was controlled by a Tesla wireless transmitter and could follow its orders even under water, for it was also a submarine!

This alone would have been sufficient to win him undying fame, had he pursued it monomaniacally. But for Tesla it was not yet enough. His idea of good transportation was a vessel which did not have to take its fuel along, but could receive it as electric power via a broadcast system. Once he had achieved the means of sending signal power, he believed that he could transmit power enough to drive ships and eventually aircraft.

Tesla had advanced in his research to the point where he was able to build a 5 million volt oscillator in his New York laboratory. He was already using lower-potential oscillators as a drive for wireless fluorescent lamps which lighted the laboratory. The big job was used to test his theories of energizing the earth to transmit power by conduction. He recognized that he had reached the limits of safe operation within the confines of the city and proceeded to set up a new laboratory near Colorado Springs. Local entrepreneur Leonard E. Curtis had promised land and all the electric power he needed, reasoning that a great scientific name such as Tesla's would bring a good deal of attention to the community.

The new laboratory was at an ideal location. Colorado terrain is a great producer of natural lightning and Tesla was able to devise equipment which would record the magnitude of the blasts. He determined that standing waves were set up by the lightning bolts – as storms moved further away, the recorder invariably showed peaks and troughs in the received energy. This further substantiated his theory of earth conduction. The additional data he obtained aided his production of artificial lightning, perhaps

The first guided missile. Tesla sent crewless boats into the Atlantic and then called them back by radio control. He further proposed self-thinking machines that would obey commands, no matter how complex, even though given far in advance. The development of the guided missile and the computer stand as remarkable examples of Tesla's prophetic vision.

with a view to controlling the weather, as he had fancied long before. However, his present interest was the development of practical wireless transmission of power in commercial quantities. Tesla's apparatus depended for its effect not on electromagnetic radiation, as some have mistakenly believed, but on massive electrostatic action. Indeed electromagnetic dispersal of power was wasteful, in Tesla's opinion. His system used a "magnifying transmitter," which was a gigantic Tesla coil, driven by thousands of horsepower. The secondary was a quarter-wavelength coil, one end grounded, the aerial portion attached to an immense copper electrode. Once the generator attained full power, the aerial electrode acted as one plate of an artificial condenser, the ground point being the other. In this way, Tesla pumped enormous quantities of electricity into the earth. Some 200 kW was the rating of his Colorado installation.

While weeks of labor turned into months, and the barn-like building he had had erected became stuffed with equipment, his workers had no clear idea what exactly was to happen. At last all was complete and Tesla explained to them; there was a nervous exodus for the door. Only one old assistant dared to stay. Tesla stood outside, conveying his orders to Czito, the assistant, by hand signals. At first, the generator, which drew its power from the commercial ac mains, was only energized a little. Then the power was slowly increased. With each step up in power, changes occurred at the copper aerial electrode. First there was only a violet corona discharge. Then, as more power was fed to the system, sparks began to snap and hiss. At the half-power point, bolts of raw electricity leapt from the electrode. As the system neared full power, artificial bolts of lightning, thick as a man's arm, snapped forth, fifty, a hundred, then almost two hundred feet in length. This secondary effect was natural, a sign that the electrostatic equilibrium was being disturbed. When the system stabilized, the lightning would end. Then, as they reached full power, all activity ceased. Tesla ran inside, saw that he was no longer receiving power from the commercial lines, and called the power company.

"You've ruined my experiment," he told them.

"And your experiment has ruined our station," was the reply. "You overloaded the generator and it caught fire."

Humbled, Tesla took Czito to help him set things right at the power station.

It has been stated that Tesla failed in his attempts to transmit power, but this writer believes that other factors became predominant. There are pictures extant of Tesla's apparatus in operation. In one instance, he demonstrated for the press the new system. A receiving circuit was set up twenty-six miles away from the "magnifying transmitter" and 200 50-watt incandescent bulbs were lighted to full power; this, with the transmitter at a low output. Tesla claimed a 95% efficiency for the new equipment.

Tesla was fully satisfied with his results and evidently the U.S. Patent Office concurred, as they issued a number of patents on wireless power transmission. However, when Tesla returned to New York to raise capital for a commercial system, he was unable to. Investors could not see what profit they could make from a system where anybody anywhere could use the power fed into the earth. Let it be said that practical financial considerations of this type were never Tesla's strong suit. The single greatest flaw of this scientific Jove was complete lack of a hard-headed business sense. He further compounded his errors by an overly generous nature.

Those who infringed on his patents were never attacked by him personally, as he considered time taken in such undertakings as wasted. He believed he could never run out of new ideas and inventions. He preferred the wealth of discovery in the laboratory. Also, his great goodwill led him to the naive expectation that, when he had produced a means of improving the human condition, he would be appropriately rewarded by a grateful humanity. His polyphase power distribution system had already improved man's lot by substituting cheap electrical power for manual work in many areas. His new system would make electrical power available in the most remote, inaccessible and impoverished places.

Tesla, returning to New York City in the summer of 1900, found himself in the same situation he had been in in 1882 – he had a wondrous discovery, but could not interest anyone in it. The financiers asked if he had not something else they could capitalize on. He stated that he could adapt his system as a "world broadcast plan." With it, he asserted, he could provide "interconnection of existing telegraph and telephone exchanges *all over the world*; establish a secret and non-interferable government communication service; maintain universal distribution of news; establish a world-wide system of musical distribution, maintain accurate time signals to clocks everywhere; provide full facsimile transmission; send accurate navigational signals" – among other possibilities.

"A cheap and simple device, which might be carried in one's pocket, may then be set up somewhere on sea or land, and it will record the world's news or such special messages as may be intended for it," wrote Tesla, concerning his "World-system."

The men of money were skeptical; Tesla promised much. Still, he had delivered on his claims before; the polyphase ac system in general use now was silent testimony to this man's genius. Then J. P. Morgan made a gift to Tesla of a sum which was never disclosed, but estimated to be from $150,000 to twice that amount. James S. Warden, manager of the Suffolk County Land Company, made available to Tesla two hundred acres of his company's tract at Shoreham, Long Island, about sixty miles from New York City.

A large brick building was soon erected and a great, wooden tower was being built. An all-wood structure was specified, because of the enormous voltages the plant would handle. Sanford White, the eminent architect of government buildings of that era, provided the plans for such a structure. The tower was some 187 feet high, and many people snickered, because they knew such a wooden tower could not stand, or if it did, would not be able to resist high winds. A torus-shaped copper electrode, 100 feet in diameter, was to top the tower. This was later changed to a hemispherical shape. Needless to say, the equipment being

A huge 187-foot tower was constructed on Long Island as part of the inventor's "World System." Not only was it to provide an incredible number of services in communications but was intended as a model plant for wireless power. Tesla claimed that it would send electricity through the earth to run the street cars of London and light the bulbs of Paris. The project was never completed and the plant remained but a landmark until it was demolished.

installed – most made to order – the tower itself, and the cost of the copper electrode and changes engendered, took all the money Tesla had. The smaller investors stood by, waiting to take their cue from Morgan. That financial baron had made a good deal of money on the General Electric Company, whose financing he had accomplished, and the others reasoned that if he was backing Tesla again, they would follow. They did not know that the sum was a personal gift, not a business investment. When Tesla ran into financial difficulties and Morgan did not rescue him, rumors arose that Morgan had withdrawn his support in dissatisfaction. Tesla was unable to raise further capital, and the "World-system" went down the drain. A couple of personal friends extended sums, which Tesla used to pay off creditors.

It may be argued that the system was a technological failure, that Tesla was unable to produce the results he had claimed. However, his earlier experiments in Colorado had been sufficiently practical to win him U.S. patents. It is hardly necessary to say that the Patent Office does not grant patents on conundrums such as perpetual motion machines.

Tesla wrote of this failure: "My project was retarded by the laws of nature. The world was not prepared for it. It was too far ahead of time. But the same laws will prevail in the end and make it a triumphal success."

Nevertheless, Tesla did not regain his stature as a creator after this episode and his twilight years were taken up with ever smaller practical projects and ever more grand dreams and theories. His next major enterprise was the successful development of a bladeless turbine, another revolutionary idea. The Tesla turbine used a rotor composed of a series of smooth discs. Steam entered through center porting and flowed in spiral lines around the discs, dragging them along by virtue of viscosity and adhesion. The first experimental model, built in 1906, was six inches in its largest measurement, weighed about 10 pounds, and developed 30 horsepower. But it was another case of being too much ahead of time and revolutionary. Though the high cost of machining the conventional turbine bladed rotor made the Tesla approach attractive from a cost viewpoint, the conventional turbine was already highly developed and accepted.

In 1911, Tesla adapted the radio-control apparatus he had developed for his 1898 submarine demonstration to an airplane and presented his plans to the War Department, where the scheme was laughed at. Of course, this was the same period when Robert Goddard, the American father of modern rocketry, was ridiculed as a crank by the military savants of this country. By 1936, the aging Tesla had completely recanted in his idea that devestating weapons would make mankind turn away from war; he claimed that he had conceived a 'death ray,' but would not give his calculations to the government. Fantasy, some say. Yet it is a fact that the Federal Bureau of Investigation impounded his papers when he died in 1943, and for all this writer can discover, still has them. The death ray would have been only a defensive weapon, because it required staggering amounts of electrical power. It would have secured a country against enemy air attack. As early as 1917, Tesla had thought about this problem and had described the possibility of apparatus which could broadcast short-wave impulses and receive reflected waves which would be displayed on a fluorescent screen – radar, in short.

Fortunately, Tesla wrote a good deal in later years. Gernsback's *"Electrical Experimenter"* published a series by the scientist starting in 1919, which gives some valid biographical data and a good review of his discoveries and plans. Tesla also wrote on diverse subjects: van de Graaf generators, Servian poets and translations of their works, the compass, the moon's rotation, woman's role in future society. Perhaps the writing that caused the greatest reaction from the public was "The Problem of Increasing Human Energy," in *Century* magazine, June, 1900. Instead of a dry scientific dissertation, Tesla took a philosophic approach, derived in part at that time of a mechanistic view of man. In this massive article, he discussed such topics as public health, morals, diet, the outlawing of war, his telautomatics and their potential use, harnessing solar energy, the iron industry, the coming age of aluminum, new power sources and prime movers, wireless and the secret of tuning, and the practical possibility of interplanetary communication.

Concerning the latter, Tesla had claimed that he had received radio-type signals which could only have come from outer space. Conventional people, both scientists and laymen, laughed. Marconi claimed the same results. Many years later, Project OZMA's radio telescope began probing the stars for intelligent signals. And the recent International Astronautical Congress in Rome heard a paper which reviewed the findings of Tesla, Marconi and the Scot, David Todd.

In 1912 the Nobel prize in physics was awarded jointly to Tesla and Edision. Though he could have sorely used the $20,000 prize money, he refused the honor. Tesla called himself a discoverer and Edison an inventor. "Placing the two in the same category would completely destroy all sense of the relative value of the two accomplishments," he wrote. By a great irony, a few years later friends persuaded Tesla to accept an honor presented by the American Institute of Electrical Engineers – the Edision medal. In 1936, the Yugoslav government awarded an honorarium of $7,200 a year to the aging scientist.

On January 7, 1943, Tesla passed away, ignored by the world he had helped to energize. He had once written, "The opinion of the world does not affect me. I have placed as the real values on my life what follows when I am dead." A prophetic statement – for less than a year after his death, the United States Supreme Court rendered a decision that the Marconi radio patents were invalid, on the basis of prior work by Tesla and others. Had this significant decision been reached during Tesla's liftime, he might not be all but forgotten today. Every school child knows of Edison. Most people have heard of Marconi – but the name "Tesla" raises blank stares. Of course, much has been written about Tesla concerning his eccentricities, his strange beliefs and habits – sensation-mongering that has led attention away from his accomplishments. However, the summation must be that his was a brilliant career – he led the way for the world during the peak of his abilities, and as his influence ebbed, he pointed to future realizations for others. His successes are beyond dispute. His failures may yet yield practical results.

Dr. Mahlon Loomi—
Marconi Followed in His Footsteps

Lily B. Rozar

An unsung rival of Marconi said, "I know that I am by some, even many, regarded as a crank' – by some perhaps as a fool – for allowing myself, to the sacrifice of material advantages, to abandon a lucrative profession and pursue this *ingis fatuus* but I know that I am right, and if the present generation live long enough their opinions will be changed – and their wonder will be that they did not perceive it before. I shall never see it perfected – but it will be, and others will have the honor of the discovery."

The man was Dr. Mahlon Loomis, and the discovery of which he spoke was radio.

Mahlon Loomis had all the elements involved in achieving success. He believed in himself – a confidence planted and nurtured by his mother and father (who was himself associated with Ben Pierce of Harvard in founding the *American Ephemeris* and *Nautical Almanac*).

Mahlon Loomis had a stern code of ethics, as well as faith, courage, and fierce energy. Aside from these characteristics, he was blessed with an inventive mind and creative talent.

Seemingly, Loomis had all the attributes fundamental in achieving success. Why had he failed where Marconi had succeeded in developing the system of the wireless? It wasn't that he didn't have logical reasoning. He could weigh alternatives and make decisions. It wasn't that he lacked the boldness one needs to succeed. This dynamic element revealed itself in his wireless experiments – and his optimism. Neither did he fail because he was lacking in concern for others. His brother said after Dr. Loomis died, "He wanted mankind to enjoy the fruits of his discoveries, maintaining that it would be the means of establishing a brotherhood among the nations and races that nothing else could accomplish; and would give to the children of men grander and truer conceptions of Deity, than now prevail."

Why, then, was he called a dreamer whereas Marconi was called a genius? The answer lies in *timing*. His ideas were born out of time. Without the leaders of his country to aid him financially, his experiments had to wait thirty years to be proven a success. The men who could have aided him lacked alertness, imagination, and vision.

Mahlon Loomis was born July 21, 1826 in Oppenheim, New York. He moved to Virginia with his family in the 1840's. His education was limited to what he learned from his father and grandfather – and from the libraries of these cultured men. Though Mahlon's education was not formal, it was complete for his time. While teaching in Cleveland, Ohio, he found time to attend a school of dentistry. Later he toured the neighborhood counties and earned fifty dollars a month practicing dentistry. In 1854 Dr. Loomis invented and patented a mineral-plate (laolin) process for making dentures. This process in dentistry was also patented in England.

It was about this time that Dr. Loomis attended a series of lectures on electricity by Professor Lovering, at Lowell Institute in Boston. Soon afterward, in 1860, he applied his time to static electricity. He experimented with the growing of plants by burying metal plates connected to batteries. But he couldn't force the growth of the plants by this process. He dispensed with the batteries and tried to use static electricity . . . by means of kites carrying metal wires. This experiment failed but Loomis did learn that electrical charges could be obtained from the atmosphere.

In 1860, Dr. Loomis – now married and head of a large family – moved to Washington, D.C. Here he practiced dentistry whenever he wasn't experimenting in electricity. Loomis was determined to find the key to wireless communication. He had sent communications over a wire between South Braintree and Falls River Station, Massachusetts – a distance of over forty miles. This was done by the aid of his atmosphere battery. His wife had hoped he would give up his crazy experiments after moving to Washington. But he couldn't give up something in which he had so much faith.

Disgusted with the mad dreamer, his wife deserted him. Soon after, he moved to Terra Alta, West Virginia, and lived with his brother – but he continued his experiments.

In 1868, in the presence of members of Congress and eminent scientists, Dr. Loomis carried on two-way "wireless" communication for a distance of eighteen miles between two mountain peaks in Virginia. Kites attached to wires and connected to the ground through galvanometers were set up on each peak.

Through this experiment true radio waves were sent out . . . this was the first time such signals had been transmitted over a distance without wires. Dr. Loomis interested a group of Boston capitalists in his discovery. He had proven to them and to those who witnessed the experiment that telegraphy without wires could be a cheap and convenient means of communication. He cited the fact that this new communication system would require no repairing of wires and it would be much safer. The men with money were convinced and ready to invest. But before transactions could be made, "Black Friday" hit the nation – it was the infamous "panic of 1869."

In the meantime, Dr. Loomis had turned to Congress for fifty thousand dollars so that he could continue further experimentation. A Senator Summer of Massachusetts did introduce a bill in behalf of Dr. Loomis' petition for financial aid. For some reason the bill was relegated to the Committee on Patents instead of the Committee on Appropriations – and no action was taken during the congressional session.

In 1870, Loomis communicated between two ships two miles apart on Chesapeake Bay – convincing himself and others that his discovery was practicable. But time passed and no help could be obtained. Without money Dr. Loomis could not develop his discovery.

Finally, in 1870, the bill incorporating the Loomis Aerial Telegraph Company was introduced into Congress. It passed the House in May 1872, and the Senate in January, 1873. It was signed by President Grant. But it failed to provide the appropriation of the fifty thousand dollars so badly needed by Loomis.

He had courageously struggled for his patent and for this charter. He thought it would be an effective means through which he could secure money for the promotion of his work. Neither his patent nor the charter proved to be anything more than a piece of paper.

Aside from his patents in dentistry and in the electrical field, Dr. Loomis invented and patented a convertible valise, and cuff-collar fastener.

On the record, Dr. Loomis is credited with the following:

First patent for a system of wireless.
First demonstration of wireless signaling.
First to employ a vertical antenna and ground.
First to employ spark signaling.
First to employ a kite for antenna support.
First to specify an "indicator" in his receiving system.
First to recognize the value of electrical agreement between his sending and his receiving system (now called "resonance").

Dr. Mahlon Loomis spent his life striking for the summit. And, though he achieved the above-mentioned firsts, he was not accepted as a good risk.

As years passed by, Loomis became known as the prophet without honor but time has proven that "prophet without honor" is another way of saying "unacclaimed success."

Fascinating Fundamentals—Volta and His Pile

W. Edmund Hood W2FEZ

Although this is about Volta and the invention of the electric cell, it would be incomplete without at least a mention of another Italian, Luigi Galvani. Both these men stumbled over the same principle, and both men misinterpreted it, although Volta's interpretation was most nearly accurate. Galvani's main field of operation was electric detectors. After his wife was scared out of her wits by a pair of frog's legs that jumped without having the rest of the frog attached, and after she suggested to her husband that the phenomenon just might have something to do with his electric machines, Galvani "discovered" this very sensitive method of electric detection.

Galvani worked for a number of years with his frog legs, and during the course of his experimenting, he completed a circuit to a pair of legs by way of a copper wire and an iron fence. He observed that the legs would jump when connected through two different metals, but not when only one kind of metal was involved. Not being

Table . When immersed in any of the solutions shown, any two of these metals will produce a voltage equal to the difference of their potentials. For instance, Zinc and Silver in a solution of lye: E = 0.958 (−0.321), or 1.719 volts.

Solution	Sulphuric Acid	Lye	Salt
Metal			
Zinc	0.0	−0.321	?
Lead	0.513	0.318	0.512
Tin	0.513	0.002	0.503
Copper	1.007	0.802	0.809
Silver	1.213	0.958	1.013

concerned with electric generation, he continued his experiments in the field of detection.

This was in the latter part of the eighteenth century, and in those days, the only source of electric power was from electrostatic machines. Just a few decades before, Franklin had discovered atmospheric electricity through his famous kite, but this had proven to be a difficult source to control. Nobody had, as yet, said so, but the thing most needed to further the progress of electricity was a source of continuous current.

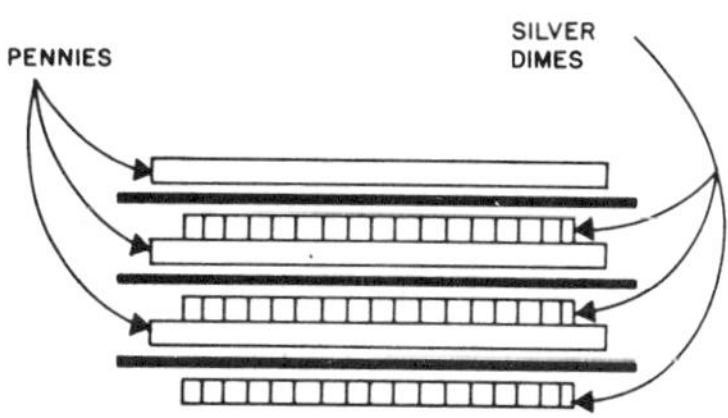

Fig. 1. Volta's pile (also called the spit-on-a dime battery). Blotters should be moistened. This always used to be made with the older silver dimes. I don't know how well it will work with the new ones.

Galvani had been convinced that his connection of the frog legs to the metal strips had shown evidence of what he called "Animal Electricity." Volta thought differently. He felt sure that the source was not in the animal tissues, but rather in the connection between the two metals. He came awfully close. For quite a while he experimented with a great many combinations of metals, and eventually came up with a list so arranged that any metal shown would generate a positive charge when contacted with any metal below it. His one misconception, and it was a minor one, was that it was the CONTACT between the two pieces that produced the charge.

Volta had tried to improve the connection between the metals by moistening them with brine. In time he found that the same result could be had if the two pieces were altogether separated. He still believed, however, that it was only the connection between the metals, and not a reaction with the brine that produced electricity. Soon he

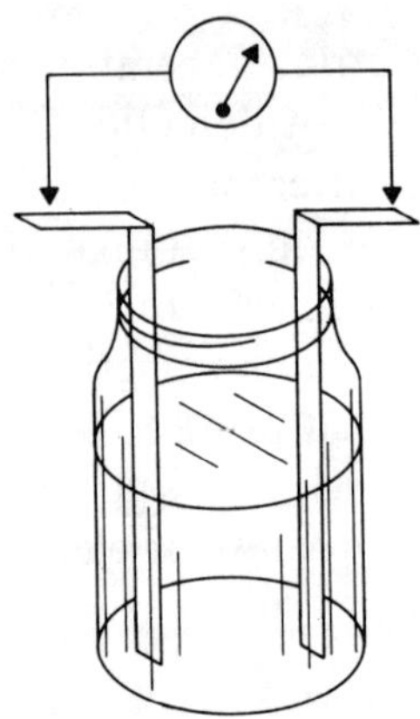

Fig. 2. A simple jar cell. See table for solution and metals.

was making stacks of his cells piled one on top of the other. This, naturally, produced a more powerful charge and, while its power was nowhere near that of the static machines, it never had to be recharged. Here, then, was a source of continuous current. Volta published his conclusions in a letter which was read to the Royal Society in England on June 26, 1800. On that day, electricity took a giant step forward. In the next half century, many improved cells were developed by such men as Faraday and Davy, but it was Volta who started it.

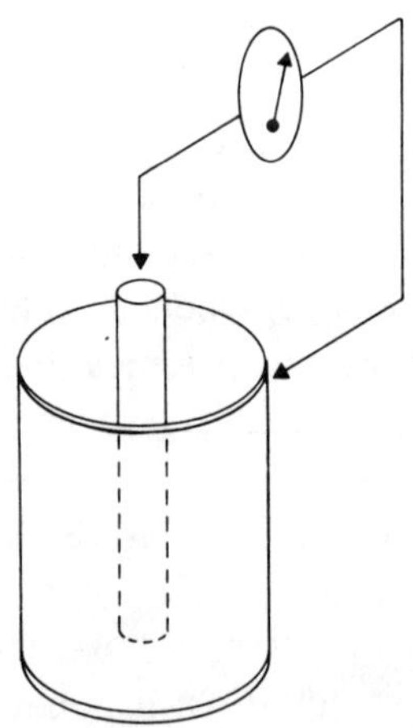

Fig. 3. The "can-full-of-vinegar" cell.

Volta's one misconception, that electricity was being generated by the action of the two metals with each other, was not too far off. Actually, it is the action of the two metals AND A CHEMICAL. In the table, we see the potentials which can be produced with a few common metals and various solutions. To determine the potential you will get, take the DIFFERENCE between the potentials of the two metals you will use. For instance, with a sulfuric acid solution, zinc has a potential of 0.0, and copper, 1.007. If you immerse a strip of copper and a strip of zinc in a jar of sulfuric acid, there will be a

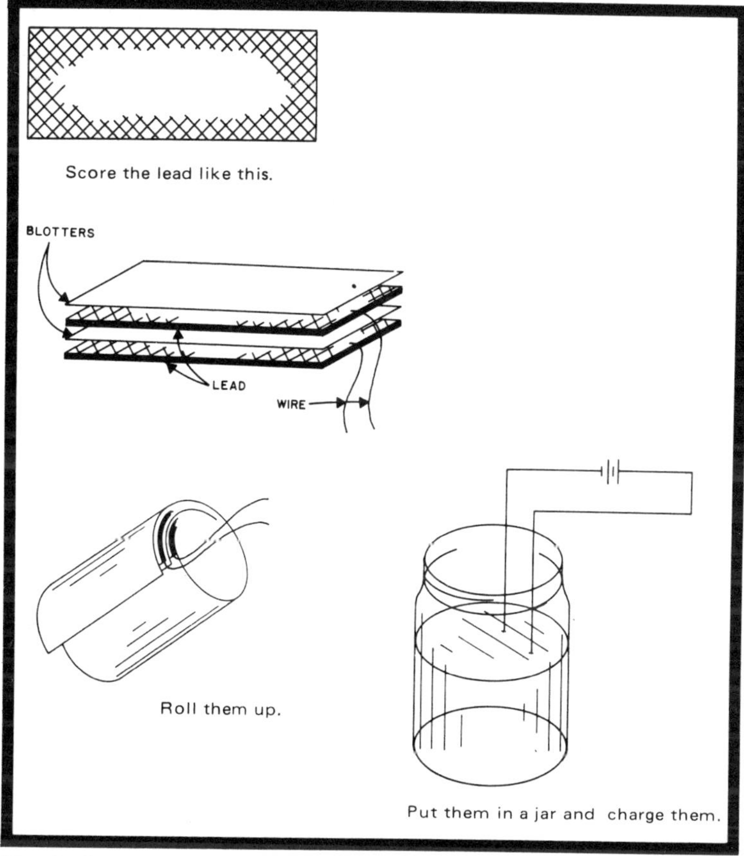

Fig. 4. A simple storage cell. Make sure the lead pieces do not touch each other. Fill the jar with battery acid obtained from a local gas station. Once it has been charged the first time, mark the polarity and always observe it in future chargings.

potential of 1.007 volts between them. Lead, however, has a potential of 0.513. A cell made of copper and lead will produce only 1.007–0.513, or 0.494 volts. Try it yourself. All you need is a peanut butter jar (empty), some metal, and some acid. You can get the acid at any gas station. Just ask for some battery acid.

I once made a demonstration cell by filling a tin can with a solution of vinegar and immersing a copper pipe. Between the copper and the tin, there was about half a volt, and if I had used sulfuric acid instead of the vinegar, that would agree quite well with the table.

A storage battery is one in which the chemicals can be renewed by sending a current through it in the reverse direction. Take two strips of lead, and score criss-cross lines on them with a scriber. Then roll them together between two pieces of blotter, making sure that they do not touch each other. Put the whole thing into a jar and add some battery acid. Now "charge" it by connecting it to a two or three volt dc source. After an hour or so, remove the source and you will find you have made a 2 volt storage cell.

Fascinating Fundamentals— Galvani's Jumping Frog

W. Edmund Hood W2FEZ

The electrical genius sat in his study pouring over dozens of musty volumes of absolutely incomprehensible scientific dribble. Elsewhere in the house – in the kitchen to be exact – his devoted wife was busy preparing his favorite dish, frogs' legs. Suddenly she dropped her knife with an ear-piercing shriek that made the windows in the attic rattle. Galvani rushed to her side. Half hysterical, she told him that the frogs' legs had jumped, and the rest of the frog had long been in the garbage pail. Upon hearing this the genius instantly deduced that the frogs' legs were sensitive to electricity, and that electricity had been present in his wife's body.

That's the way I first read it in a third-grade storybook many years ago. It makes a cute tale, but frankly that ain't the way it happened. In more recent years, I've heard two more credible versions of the story. One said that Galvani himself saw the legs jump as he was preparing them for his wife. The other, which I am more inclined to believe, is Galvani's own version. The legs happened to be on a dish in his lab, and his assistant was diessecting them. His wife noticed that they jumped every time a nearby static-electricity machine sparked, and she suggested that they might be sensitive to electrical charges.

Well, so what?

Luigi Galvani (1737–1798) or, as he called himself, Alosio Galvani, was not originally an investigator of electricity. He was as interested in it as any other scientist, but primarily he was a professor of anatomy.

However it happened, his discovery that electrical charges could make animal tissues contract was one of the more important steps in the progress of electrical science. Its story divides and goes in three different directions. First, the direction Galvani followed:

Galvani concluded that electricity was present in animal tissue. In fact, he believed that he had stumbled onto the substance of life itself. He was partly right. Electricity is present in *living* animal tissue. His means of proving it, however, were all wet. For example, when he inserted two metal probes, made of different metals, into the frog legs, they would jump when the metals touched. Frog legs hanging on a metal fence would sometimes jump all by themselves, even if there was no static machine nearby. Galvani's countryman, Volta, vehemently challenged the first of the above experiments. He claimed that Galvani had only demonstrated that electricity could flow from one substance to another. Volta later made an electric battery by bringing two different kinds of metal into contact with a chemical. Volta was heaped with honors. Galvani died broke and broken-hearted, but he stuck to his theories to the end.

The second path of Galvani's discovery was a direct result of the first. Quite a few scientists believed Galvani's theory of "Animal Electricity," and considerable research was done in that line. In 1818, for example, an English murderer, condemned to hang, sold his body to a scientist in the hope of being returned to life. The scientist sent powerful surges of current through different parts of the body, and the results were absolutely gruesome. The body twitched, moaned, squirmed, and even breathed heavily. The hangman was not to be cheated, however. The killer stayed dead. Many a poor cripple watched hopefully as electric charges made hitherto useless limbs kick convulsively. Eventually, however, the effects of electricity on animal tissue were used beneficially. Hospitals use a Phrenic Respirator to help a patient breathe by applying electrical charges to a nerve in the neck. Many a coronary patient is alive today because of the fantastic "Pacemaker" which keeps the heart beating properly by applying precisely regulated electric pulses directly into the heart. Had he

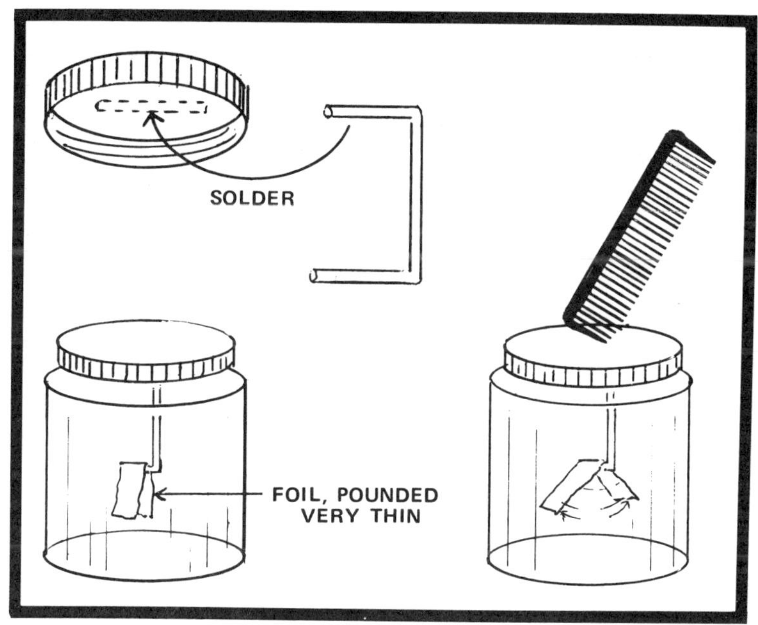

been able to foresee the future, Galvani would certainly have died happy.

Finally, the way in which his name was immortalized: For many years, frogs' legs were one of the most sensitive means of detecting electricity. Today instruments which detect minute electrical charges are called *galvanometers*.

Galvani had stumbled over the principle of Volta's electrical cell. For a long time, this device was called a *Galvanic* battery.

You could duplicate Galvani's experiments, I suppose, if you wanted to sacrifice some poor, unfortunate frog. I've never tried it, but there is no reason why it wouldn't work. One interesting experiment, however, is to make an early form of a galvanometer. All you need is a peanut butter jar (empty), some wire, and a small strip of aluminum foil. Bend a short length of bare copper wire, No. 20 or larger, as shown in the diagram and solder it to the center of the jar cover. Take a three-inch strip of aluminum foil, ½ inch wide, and pound it down until it is very thin. Hang it over the end of the wire and put the cover on the jar. Remove the paint from the top of the cover.

When the metal cover is touched with a charged object, such as a glass rod after rubbing with wool, the ends of the foil will spread. This is because the two ends have a like electrical charge and repel each other. Touch your finger to the metal cover and, with the other hand, touch a water pipe. This drains off the charge and allows the ends of the foil strip to come together.

Fascinating Fundamentals— The Terrible Jar at Leyden

W. Edmund Hood W2FEZ

The year was 1745, and scientific circles were all agog over that strange force called electricity. In England, several decades earlier, Francis Hawksbee had built an electric machine which produced unprecedented quantities of this mysterious power, and Stephen Gray had demonstrated that the power could actually flow from one substance to another. Two Frenchmen, Francois de Cisternay Dufay and Abbe Nollet had discovered two different kinds of electrical power, each capable of neutralizing the other.

Many theories had been offered as to the nature of electricity. The most popular was that it was some kind of an invisible fluid. (That's why we refer to it today as "juice.") All in all it certainly was an exciting time.

Leyden is a small town in Holland, about fifteen miles north of Rotterdam. It was there, during these exciting times, that a distinguished scientist, van Musschenbroek, set out to find a way to collect this electric fluid. Could it possibly be collected in a bottle? With the aid of a pupil named Cuneus, van Musschenbroek connected his Hawksbee electric generator through an iron chain to a gun barrel. Another chain hung from the other end of the gun barrel. Van Musschenbroek had his student hold a jar so that the loose chain dangled into it. After the machine had run for awhile, the jar was found to have an electric charge, but there was no fluid present. Van Musschenbroek then decided to put some water into the jar. He knew that electric fluid seemed to flow most easily through wet materials.

Soon they were set up and the machine was running again. Cuneus held the bottle at one end while the professor turned the generator. This time they carefully watched the level of the water, hoping to see evidence of more fluid. Time passed, and the liquid level in the jar was unchanged. The experimenters were discouraged. Van Musschenbroek stopped the machine, and Cuneus reached up to disconnect the chain.

Suddenly Cuneus dropped the jar and seemed almost to fly backward. The professor rushed over to his student. Cuneus was pale and scarcely breathing. He had received a shock such as no man had ever felt before. After his student recovered, van Musschenbroek wrote of his "terrible" experiment to a colleague at the French Academy. He advised that no one else try it. Naturally, everyone else did.

At the same time as van Musschenbroek performed this nearly fatal experiment, the Bishop of Pomerania, E. G. von Kleist, did essentially the same thing with similar results. He failed to publish his findings through acceptable channels, that is, by writing of them to a university or other seat of learning, so the credit went to van Musschenbroek. The device was named the "Leyden Jar."

The scientists didn't realize it at the time, but they had actually made a capacitor. The water served as one conductor and the student's hand as the other. The glass was the dielectric. It was later to be improved by using foil conductors, but in the meantime the water was thought to be essential. The scientists believed that the cool water condensed the electric fluid, and the Leyden Jar was nicknamed a "condenser." That term is still used to this day.

The home experimenter can make his own Ledyen Jar quite easily. All that is needed is an empty peanut butter jar, some aluminum foil, and a small length of chain. The inside and outside of the jar are lined with foil up to within ½ in. of the top. The chain is soldered to the center of the cover, and hangs down into the jar. It should be long enough so that a couple of inches rests on the foil on the bottom.

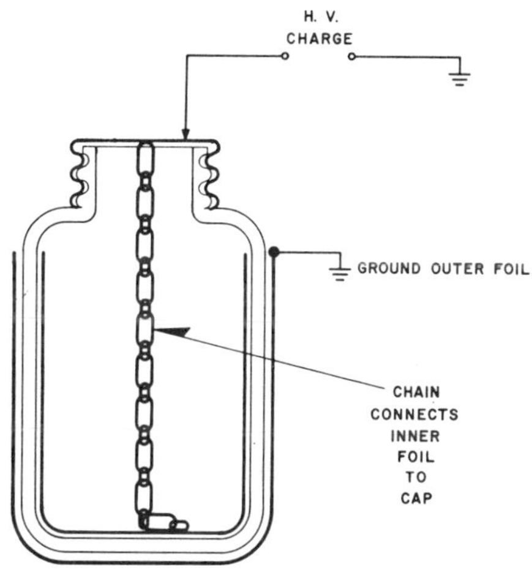

To charge your Leyden Jar, the outside foil should be grounded. You can use a battery, high voltage supply, or you can charge it up by static electricity. Simply rub a glass rod with a piece of silk and touch the rod to the cover of the jar. If you do this enough times, the jar will accumulate quite a charge.

Now, since some clod might be stupid enough to try it, *don't* connect it to a lightning rod. The early scientists who tried that seldom lived long enough to publish their results.

I once made a capacitor to demonstrate the principle to a group of Boy Scouts, using a roll of waxed paper and a roll or two of aluminum foil. Begin by unrolling four or

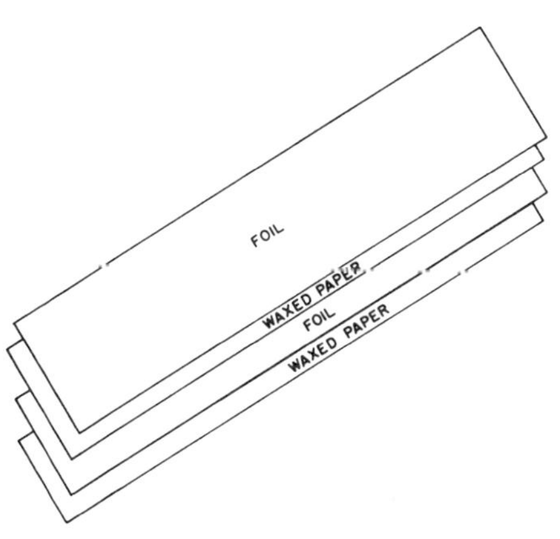

five yards of waxed paper onto the floor. Place a strip of aluminum foil on top of this so that the foil overlaps the paper by a few inches on one side, and the paper sticks out an inch or so on the other side. Lay a strip of waxed paper

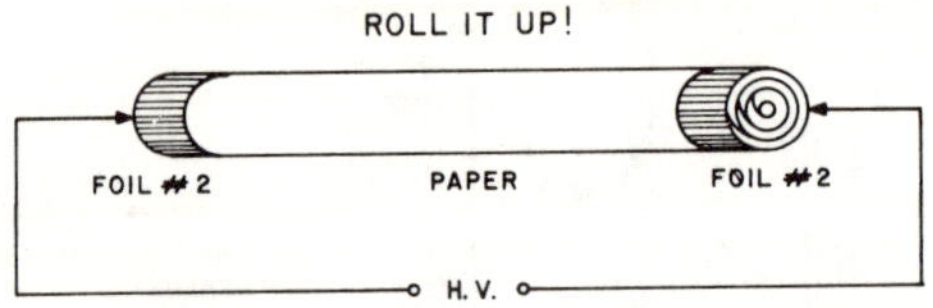

on top of this, exactly covering the first. You now should have a sandwich with the foil in the middle, one edge sticking out. Lay another strip of foil over this whole thing, but protruding on the other edge. The ends of the waxed paper should stick out far enough so that the pieces of foil do not touch. Roll the whole mess up as tightly as you can. If you do it right, you'll have a fairly good capacitor. It should take a charge up to 100 volts or so. Mine broke down at 150.

Fascinating Fundamentals—Electrostatics

W. Edmund Hood W2FEZ

No one can really say who the discoverer of electricity was. Long before the birth of Christ, the Greeks were aware of the mysterious properties of Amber. Amber is a glass-like clear yellow substance. Once the resin of ancient pines, it has been slowly petrified by the ages. The early peoples noticed that, when briskly rubbed, a piece of amber would draw tiny bits of hair and dust. As if hurled by some unseen force, they would suddenly leap up and cling to the amber. Then, just as suddenly, they would fly away.

For centuries, this phenomenom remained a petty curiosity. Then, as the knowledge of mankind advanced, it was destined to be the key to man's most miraculous achievements. The Greeks called amber ELEKTRON. From this modest beginning comes electricity, electronics, and the name of the real culprit, the electron.

We know today that the electron is that tiny bit of stuff that orbits around the nucleus of an atom, but this was far too advanced for the early pioneers of electrical science. Hampered as they were by fear and superstition, their progress was painfully slow. For example, one of the earliest observers of the properties of amber was Roger Bacon (born 1214). Although he was at first encouraged by an enlightened Pope, he was eventually branded as a blasphemer and cast into prison. He came out in 1291, an old man.

Bacon's observations were the first really clear ones. But it was not until 1600 or so that William Gilbert, a private physician to Queen Elizabeth I, began to introduce some light on the matter. Gilbert discovered that many

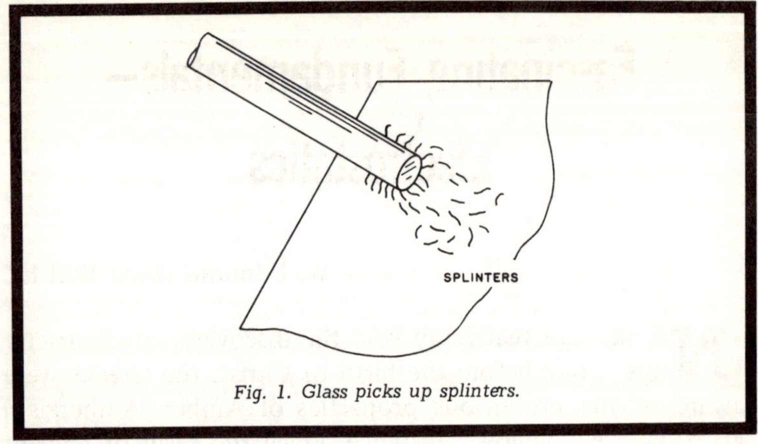

Fig. 1. Glass picks up splinters.

substances showed these properties. He called them "electrics." The force he called "vis electrica." He even made a crude instrument for measuring the electrical attraction – the "versorium." His work was widely published, and drew praise from the great Galileo.

None of these early pioneers, however, could understand or discover the reasons for the phenomena they observed. They had no precedents to refer back to, and even the best of their tools were clumsy by our standards. They had to grope their way along. We are fortunate in that we can backtrack on the trails of progress and examine their discoveries in the light of modern knowledge.

The secret lies hidden within the tinest of particles – the atom. An atom is itself composed of particles – particles of energy. The most readily accessible of these is the electron, the work-horse of electricity.

An atom is made of three kinds of primary particles: protons, neutrons, and electrons. The inner part consists of protons and neutrons, while the electrons revolve around this nucleus like planets around the sun. Of these three particles, only protons and electrons are of electrical importance, electrons primarily.

Let's try a simple experiment. Take a glass rod, and rub it briskly with a piece of silk. Now hold it close to a few splinters. The splinters will leap up to the glass, cling to it for a few moments, then fly away.

When you rubbed the glass, you actually wiped electrons off the atoms on the surface. (When there is a shortage of electrons, we call it a POSITIVE charge.) Any electrically charged body will attract uncharged particles. As the splinters clung to the glass, it drew electrons away from them. Soon they also had a positive charge. Two objects with the same electrical charge will repel, or push each other away.

Take a piece of hard rubber and rub it briskly with a piece of wool or simply draw a hard rubber comb through your hair. This rubs electrons onto the rubber, making a NEGATIVE charge. A toy balloon rubbed on your shirt sleeve will stick to the wall, because it becomes charged.

There are a great many materials which you can charge by rubbing. A piece of paper will cling to the wall after being rubbed with a wooden stick. Your clothing, rubbing against the seat of a chair, or an auto seat, will generate enough electricity to give you a snappy shock!

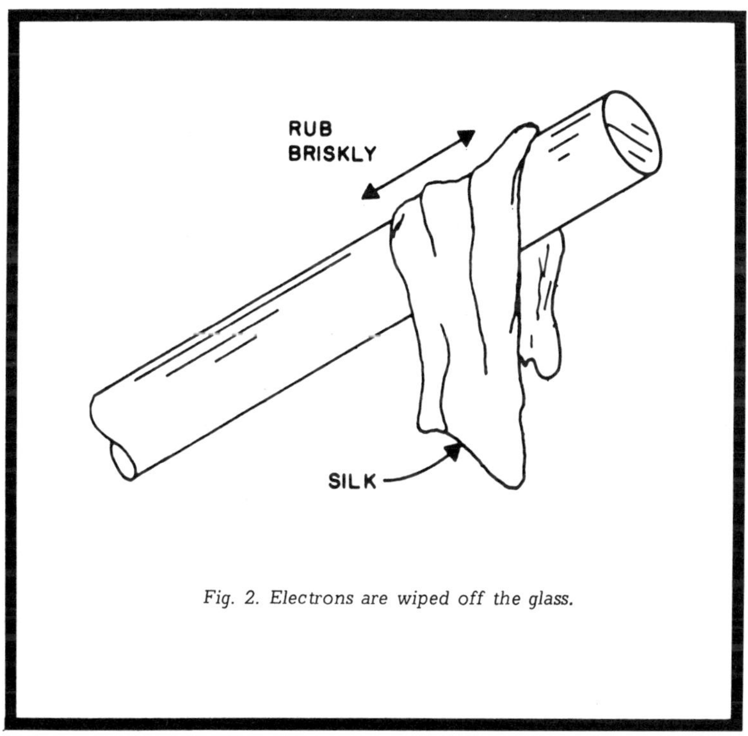

Fig. 2. Electrons are wiped off the glass.

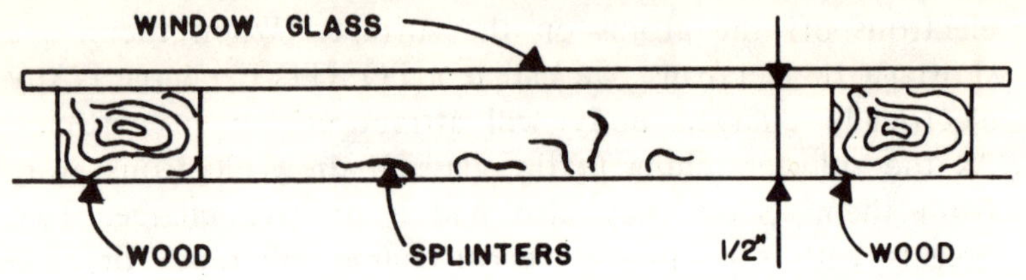

Fig. 3. Wipe the glass and watch the splinters dance.

Here is an easy-to-make demonstration. Sprinkle a few splinters on a table-top. Place a piece of glass about ½ in. over them. Wipe the glass briskly with a silk cloth. and watch the splinters dance! This works best on a dry day; the dryer, the better. On a rainy day it may not work at all.

Fascinating Fundamentals—Magnetism
The Mysterious Lodestone

W. Edmund Hood W2FEZ

It was a balmy spring day in the year 435 B.C. A sturdy ship made its way slowly across the water. The captain stopped pacing the deck and looked sleepily out at the towering mountains on the nearby island. Suddenly there was an ominous creak. Something flicked by the captain's ear, drawing a little blood as it went. Puzzled, he looked around. He opened his mouth in astonishment at the sight of three shields sailing through the air toward the distant island. A warning shout from behind, and the captain hit the deck as a sword whipped past his head. Looking up cautiously, he gasped as he beheld planks which had fallen from his ship, bobbing in the wake. The awful truth dawned on him. He had sailed into the grip of the terrible Lodestone Mountain.

That looks like the start of a king-sized fish story, but the fact is, sailors of old actually believed such a mountain existed. It was one of the calculated risks of going to sea. So great were the magnetic powers of the fabled mountain that the nails would be drawn from passing ships, and they would disintegrate! This was typical of the legends concerning lodestone and magnetism in the days of yore.

Just when magnets were discovered, and by whom, is lost in antiquity. While the Chinese are known to have used magnetic compasses around 300 A.D., wild and wonderful tales of this mysterious force circulated around both the eastern and western worlds centuries earlier.

For instance, there was Magnes, a shepherd boy whose nailed boots and iron tipped spear were clamped to the ground. His name is thought to be the origin of the word magnet.

Or was it derived from Magnesia, a place where great quantities of this magical stone were found? Who knows?

For a long time, though, lodestone, a naturally magnetized mineral, was shrouded in mystery. In the west, the compass came into widespread use around the turn of the 13th century, but superstitition still took priority over knowledge. It was believed that garlic would interfere with the compass. (I wonder how Marco Polo survived!) Also, while scientists agreed that the earth was flat, they knew that it had magnetic properties, and magnetic poles.

In 1269, Peter Peregrinus used a magnetized needle to discover the concentrated areas of power in a chunk of lodestone. William Gilbert, in 1600, confirmed Peregrinus' findings, and pressed the theory that the earth itself was a large magnet.

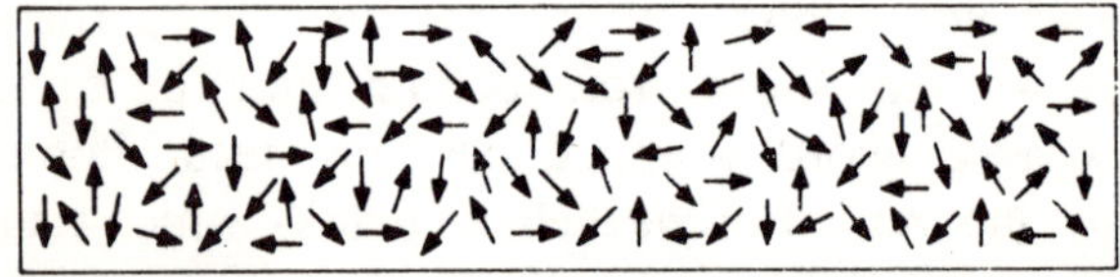

Fig. 1. When a piece of metal is NOT magnetized, the molecules all point in different directions.

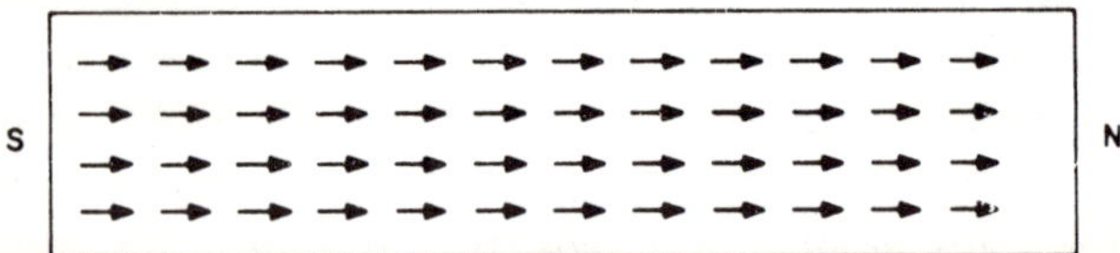

Fig. 2. Magnetizing causes the molecules to line up so that all their magnetic fields combine.

Today there is still a lot that is unknown about magnetism, but scientists nonetheless have a pretty good idea as to how it works. There are only three elements which are strongly affected by magnetism; namely, iron, nickle, and cobalt. All other elements are either very weakly affected or not at all. The alloy, alnico, most widely used of all magnetic substances, is a mixture of aluminum, nickle, and cobalt.

In a magnetic substance the molecules can be pictured as if they were themselves tiny magnets. Normally, they are all pointing in different directions. But when the substance is magnetized, the molecules line up together so that their magnetic fields reinforce one another. Anything that will cause a disarray of the molecules, such as jarring or heating, will weaken or destroy the magnetic powers.

You can easily magnetize a tool or other magnetic object with a magnet. Simply rub the magnet from one end of the tool to the other. Take the magnet away, and go back to the point where you began and rub the tool again. Several strokes like this, and the tool should be magnetized. There are two ways you could demagnetize it. First you could heat it, but that would destroy the temper. Better still, hold the tool within the loop of a soldering gun tip, pull the trigger, and slowly withdraw the tool. The alternating current through the tip produces an alternating magnetic field which destroys the fixed order in the tool. Try it; it works.

A magnetic field can be seen quite easily. Simply lay a sheet of paper over the magnet, and then sprinkle iron filings on the paper. The iron filings will line up in a very nice sketch of the magnetic field.

The ends of the magnet, or the points where the power is concentrated, are called the POLES. They are so called because, if you hang up a bar magnet and allow it to swing free, the ends will line up pointing north and south.

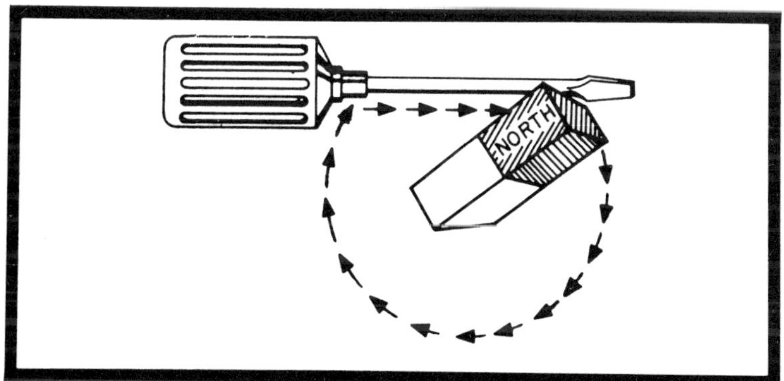

Fig. 3. Magnetic power can be transferred from the magnet to the screwdriver.

If you bring two north-seeking poles together, they will repel each other strongly. If you bring a north-seeking and a south-seeking pole together, they will attract.

The needle of a compass is really a small magnet. I remember once, as a Cub Scout, sticking a magnetized

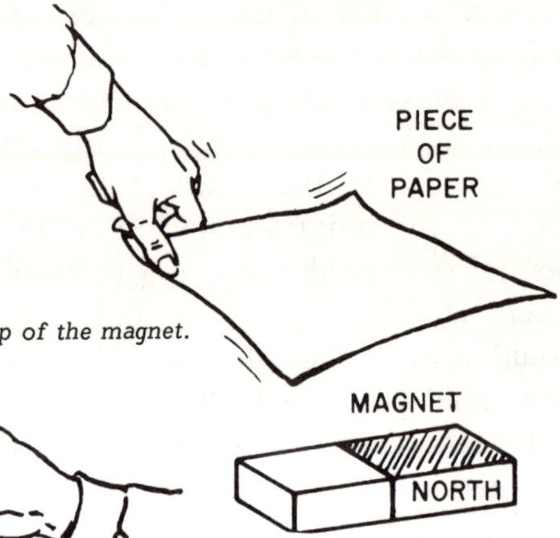

PIECE OF PAPER

1.) Lay a piece of paper on top of the magnet.

MAGNET

NORTH

2.) Sprinkle iron filings over the paper (a salt shaker is a handy way to hold the filings.)

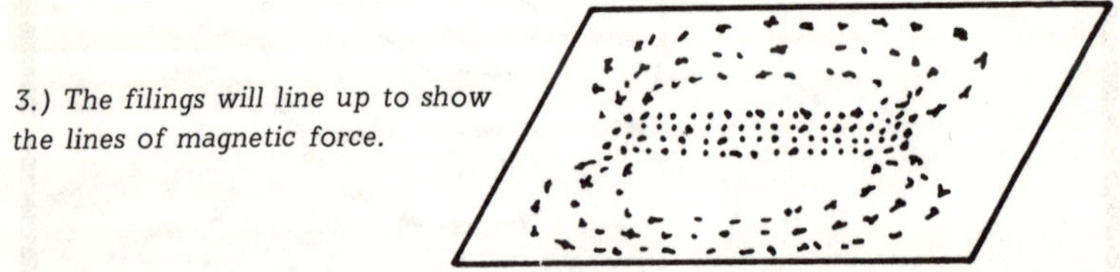

3.) The filings will line up to show the lines of magnetic force.

Fig. 4. Sketching a magnetic field.

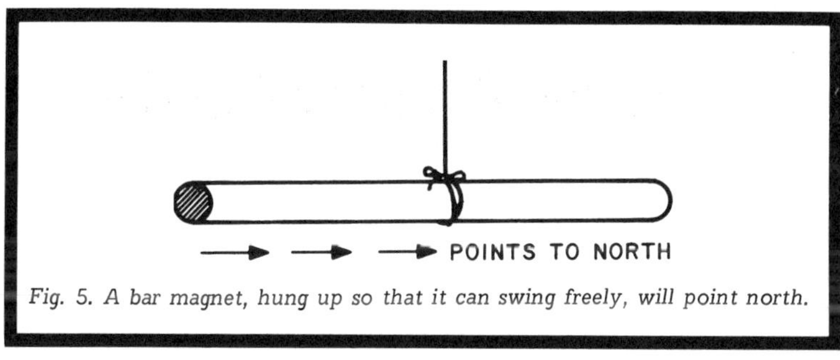

Fig. 5. A bar magnet, hung up so that it can swing freely, will point north.

needle through a cork and letting it float. Sure enough, it turned around until it pointed north.

Try this one. Take a long steel rod, point it north and touch one end to the ground. Strike the other end a few raps with a hammer. Now bring the rod close to a compass. When one end of the rod is brought near the compass, the "north" end of the needle will swing toward it. Turn the rod end-for-end, and the other end of the compass needle will point to it. That is iron-clad proof of magnetism. If you're lucky, it may even be strong enough to pick up a small pin. You will have magnetized the rod from the earth's magnetic field!

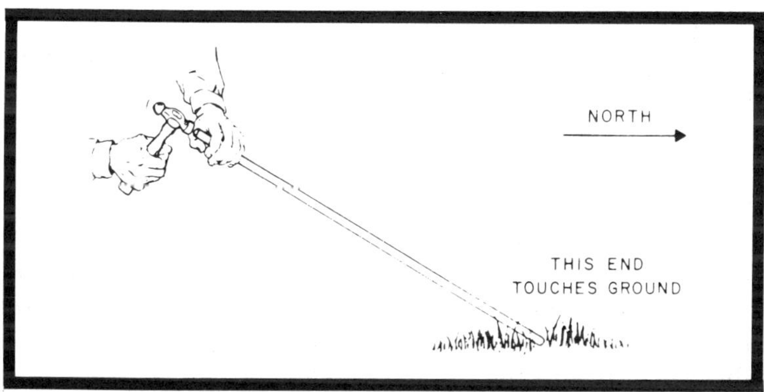

Fig. 6. To magnetize a steel bar, touch one end to the ground, point it north, and rap it a few times with a hammer.

Inside the Commercial Station

Paul Schuett

When I was chief engineer of commercial broadcast stations, from time to time someone would meekly enter the station and hesitatingly ask if he could look around. Generally these were amateurs or SWLs who were genuinely interested in knowing what was going on inside the radio station.

At most stations, you have no need to fear being inquisitive; most everyone is friendly and eager to show the visitor around. It is generally best not to schedule your visit for a Friday, since that is usually the broadcaster's busiest day. Weekends sometimes are problems too, because stations frequently operate with part-time personnel, and the person you want to see may not be around. As a rule, a Monday or Tuesday will get you the best results.

When you come in, identify yourself and explain your interest – CB, ham, or just "listener" – then just mention that you'd like to see what the operation looks like. At the larger stations, a receptionist will introduce you to the engineer or some other honcho; and from there on, you're in business. If the station has a separate transmitter, introduce yourself to the engineer on duty. Many of these separate transmitters are now remote-controlled, but the rules require a daily inspection by an engineer, so being in the right place at the right time will get you in.

The first thing you'll see in the control room will be the speech input equipment. Most stations have a commercially built audio console facing the operator. All microphones, turntables, tapes, network lines, and remote lines feed into this console; there are phone-line outputs, speaker outputs, utility outputs, connections for the on-the-air light, intercom systems, and other features. The console does all the switching, speaker muting, mixing, and amplifying. External connections are made to a terminal strip inside.

From the console, the audio signal goes through some other equipment, probably located in a nearby equipment rack. Almost every station has a limiter. This limits the level of audio peaks so that the overall signal can be fed louder into the transmitter, increasing the percentage of modulation, but eliminating the peaks overmodulating the transmitter.

Open wire transmission line has impedance around 250Ω and can be used easily with powers at 50,000W and more. Note ease of inspection and serviceability. This installation could be improved by increasing tautness of line and using metal poles. Two inside wires are hot; outside four are grounded.

In many stations, an automatic gain amplifier is ahead of the limiter. This device keeps a constant output regardless of the input level. The operator doesn't have to keep "riding gain" with one of these in the system (eyeballing the VU meter and adjusting the volume constantly). A reasonable input signal will have optimum output all the time.

A few stations employ another device to insure that negative and positive phases of the audio wave are

Transmitter and equipment rack. Note metering on transmitter – bottom meter is a multimeter with 6 switched positions. Top to bottom on rack: frequency monitor, modulation monitor, patch panel, limiter, agc amplifier, "Symmetra-Peak," homebrew test panel, spare limiter, and power switch.

symmetrical. The AGC equipment and transmitter will work more efficiently on a more symmetrical signal.

Also in equipment racks will be the station monitoring equipment. There will be a frequency monitor, indicating the deviation in hertz from the assigned frequency. Standard broadcast stations have a tolerance of 20 hertz; FM stations 2000 hertz from the carrier frequency. The meter gives a direct readout of the frequency deviation.

I always set the carrier frequency five hertz off center as a check on the monitor. For months and months the monitor would read zero deviation, but once the measurement service checked us as ten hertz high and the monitor still read zero. We fixed the monitor. With the carrier frequency a little off, a zero reading might be questioned by the alert engineer.

The station will also have a modulation monitor – a continuous reading of the percentage modulation of the signal. One meter on this instrument is for "carrier" and is

View of 250W auxiliary transmitter showing extensive metering. Every plate and grid in the transmitter is metered.

Closeup of part of equipment rack. External audio circuits terminate on patch panel so that equipment can be inserted or removed with a patch cord instantly, instead of screwdriver in a couple of minutes.

set at a reference point. This meter is also useful for checking carrier shift. The other meter reads the percentage modulation and can be switched between positive and negative peaks. The FCC requires negative peaks to be monitored; however, the switch gives the engineer valuable information about his transmitter. Nonsymmetrical peaks may mean bad modulator tubes or problems in the rf driver, etc. There is also a flashing light on this unit that can be set to fire at any predetermined percentage modulation level.

The modulation monitor generally has two outputs – one for driving the speaker monitor amplifier (this is the off-the-air signal) and the other for attaching a distortion and noise meter to make equipment measurements.

If the station operates with a directional antenna, there will also be a phase monitor, reading the relative currents in each tower and the phase between the towers.

The transmitting equipment will be probably the most interesting for the amateur. The actual transmitter will be considerably larger than the amateur's, but consider that it is designed for 24-hour continuous operation, should run cooler, and has many features not found on a five-band transceiver. The metering is more extensive – the FCC requires final plate voltage and current to be metered, along with antenna current. In addition. there will probably be a filament voltage meter, final grid current meter, meters on the modulator plates, driver plate, oscillator plate, etc. Some of the tubes used will be familiar to the amateur – 4-125A and 833A are common in low and medium power operations. My favorite tube is the 892R – a big heavy thing that isn't designed in new equipment any more. Most of the tube is copper cooling fins, with a large glass bottle coming out of the top. It stands about three feet and weighs in around 50 pounds or so. This is found in older 5000 and 10,000 watt transmitters.

If the station is using a directional antenna, you will find phasing and branching equipment, used to split the rf among the towers into the proper current and phase ratios. These have considerable plumbing inside (copper tubing usually used) with many coils and capacitors.

Look at the transmission line going out to the antenna. RG8/U is popular with lower power operations; FM probably uses copper coax with pressurized lines. Older stations may use open wire line.

At the tower, you will find a tuning unit to match the impedance of the line to the impedance of the tower.

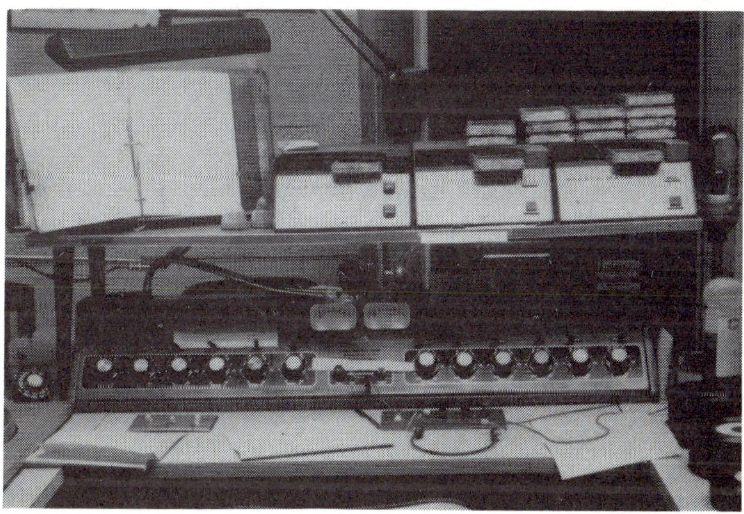

Announcer's eye view of two typical broadcast installations. Operating these is as automatic as driving a car, once you're used to them.

There will be an rf ammeter in this equipment so that the antenna current can be read. There probably will be a diode or some other device to send a relative power reading back to the transmitter building so that the operator doesn't have to hustle out to the tower every half hour to make the required meter reading for the log.

Announcer at work. Transmitting equipment in rear. Note meters are visible to operator. Rack contains tape cartridges recorded with the commercials. Players are above the console. Remote control panel by announcer's hand starts cartridge machines and reel-to-reel machine, out of picture to right of cartridge rack.

Should you ask them for a job while you're there? Maybe so – a radiotelephone first class license quite probably will be required, although some stations can use operators with third class telephone licenses with broadcast endorsement. Stations where I have been frequently have used part-time personnel on weekends or evenings. In smaller stations you will find even the chief engineer works part time.

Look around at the Teletype machine delivering the news, the salesmen, the production of commercials, and be sure to ask questions. (Be careful not to talk when the mike is open.) Chances are you'll be invited back again.

WWV—Pioneer in Standards Broadcasting

Arthur Levy WA1AAU

The National Bureau of Standards radio station, WWV, began operating from its new home in Fort Collins, Colorado, early in December 1966. The station was moved from Greenbelt, Maryland, to the present site, 60 miles north of Denver, at a cost to the government of $970,000. The move was prompted by rapidly obsolescing equipment, the need for a more central location, the high ground conductivity of the new site, and the proximity to the N.B.S. frequency standard at Boulder, Colorado.

WWV's services are among the most widely used and vital services provided by the National Bureau of Standards. Its time and frequency signals are used by ships, aircraft, electronic laboratories, radio and television stations, electrical power companies, and the makers of musical instruments (who depend on WWV's tone for standard pitch).

Amateur radio operators around the world account for 35% of WWV and WWVH (Maui, Hawaii) listeners. Hams use the signal to calibrate their equipment.

WWV joins two other N.B.S. standard frequency radio stations at the Fort Collins site. The stations, WWVB and WWVL, were established in 1963. They operate on low frequencies making possible world-wide coverage.

Station WWV broadcasts on frequencies 2.5, 5, 10, 15, 20, and 25 MHz. The broadcasts are continuous, night and day, except for a four minute period each hour. The silent period commences at 45 minutes (plus 15 seconds) after each hour.

The 5, 10, and 15 MHz transmitters deliver 10 kW to the antennas, while the 2.5, 5, 20, and 25 MHz transmitters deliver 2.5 kW. The linear amplifier, which has a 40 kW input, is driven by a one watt driver. All of the antennas at WWV are vertical, half-wave dipoles, and are omnidirectional. They are fed with 50Ω 3 5/8 in. coax cable. Antenna height varies from 20 to 120 ft.

At WWV, all modulation is double sideband amplititude with 75% modulation on the steady tones and 100% on the second pulses and voice.

In case of a power failure the station is tied into two power grids in addition to having an emergency power generator.

Since December 1, 1957, the standard radio transmissions from WWV and WWVH have been held as nearly constant as possible with respect to the atomic frequency standards which constitute the United States Frequency Standard. The U.S.F.S. is maintained and operated by the Radio Standards Laboratory of the N.B.S. at Boulder, Colorado.

The frequencies transmitted by WWV are held stable to 5 parts in 10^{11} at all times, according to the National Bureau of Standards. Deviations at WWV are normally less than 1 part in 10^{11} from day to day. Changes in the propagation medium (Doppler effect, etc.) result, at times, in fluctuations in the carrier frequencies received, and may cause greater error than noted above.

Standard audio frequencies of 440 Hz and 660 Hz are broadcast on each carrier frequency at WWV and WWVH. The audio frequencies are transmitted alternately at five-minute intervals starting with 600 Hz on the hour.

The 440 Hz tone is the note A above middle C, which is the standard in the music industry throughout the world.

Universal Time (referenced to the zero meridian at Greenwich, England) is announced in International Morse Code each five minutes from WWV and WWVH. The time announcement refers to the time when the audio frequencies are resumed. The station also broadcasts a voice announcement every five minutes in Mountain Standard Time. It is given during the first half of the fifth minute and is in English.

In addition to the time signals, WWV also broadcasts

radio propagation forecasts in International Morse Code during the last half of every fifth minute of each hour. The forecast tells users the condition of the ionosphere at the time of broadcast and for the following six hours. A world-wide network of geophysical and solar observatories feed information, which includes radio soundings of the upper atmosphere and short wave reception data, to the Telecommunications Space Disturbance Center at Fort Belvoir, Virginia. The forecasts are sent at 0500, 1200, 1700, and 2300 UT. They are broadcast in Morse Code as a letter and number. The letter identifies the radio quality at the time the forecast is made. The letters denoting quality are "N," "U," and "W." They signify that the radio propagation conditions are either normal, unsettled, or disturbed. The number portion is the forecast of radio propagation quality on a typical North Atlantic path during the six hours following the forecast.

The forecasts are made for the North Atlantic area using a path from Washington, D.C. to Frankfurt, Germany, as a standard. They are used, for the most part, for direct point-to-point radio telephone transmissions. The scale used for radio quality is based on a one to nine scale which follows:

Disturbed grades (W):

1. Useless
2. Very poor
3. Poor
4. Poor-to-fair

Unsettled grade (U);

5. Fair

Normal grades (N):

6. Fair-to-good
7. Good
8. Very good
9. Excellent

Another service of WWV and WWVH is the broadcast of current geophysical alerts. The alert tells what days there will be outstanding solar or geophysical events and when these events have occurred in the past 24 hours. The broadcast is made during the first half of the 19th minute of each hour. The letters GEO are sent in CW followed by a letter repeated five times. The letters are:

M – Magnetic storm
N – Magnetic quiet
C – Cosmic ray event
E – No geoalert
S – Solar activity
Q – Solar quiet
W – Stratospheric warning

A time code is also broadcast by WWV for one minute out of each five, ten times an hour. The code provides a standard base for scientific observations. The code is transmitted at a 100 pps rate and is carried on a 1,000 Hz modulated signal. The code contains the Universal Time in seconds, minutes, hours, and day of the year. The code is synchronized with the frequency and time signals.

The time standard uses a Cesium Atomic Beam to calibrate the oscillators, dividers, and clocks, which generate the controlled frequency and N.B.S. time scales. Information from this reference is fed to receivers which monitor the transmissions from Fort Collins. The signal is compared to the reference phase. If an error exists, a signal is transmitted from Boulder to Fort Collins by a 50 MHz transmitter. Automatic correction equipment at Fort Collins corrects any error.

The oscillator controlling the transmitted frequencies and time signals is continuously compared with the LF and VLF signals. Adjustments are then made to the controlling oscillators. To assure that systematic errors do not enter into the system, the N.B.S. time scale is compared with the transmitting station clocks by the use of a very precise portable clock. By this method time synchronization to a few millionths of a second can be attained.

. . .WA1AAU

NOTES

NOTES